BREAKING THE ICE: ANTARCTICA, CLIMATE CHANGE AND ME

PROFESSOR DAVID VAUGHAN

Foreword by Professor Emily Shuckburgh

AuthorHouse™ UK
1663 Liberty Drive
Bloomington, IN 47403 USA
www.authorhouse.co.uk
UK TFN: 0800 0148641 (Toll Free inside the UK)
UK Local: 02036 956322 (+44 20 3695 6322 from outside the UK)

This book is printed on acid-free paper.

ISBN: 979-8-8230-8732-2 (sc)
ISBN: 979-8-8230-8734-6 (hc)
ISBN: 979-8-8230-8733-9 (e)

Library of Congress Control Number: 2024907870

Print information available on the last page.

Published by AuthorHouse 15 May 2024

authorHOUSE®

For Jacqui

Contents

Foreword

"The Polar Regions may be at the ends of the Earth, but what happens there affects us all. Places of compelling beauty and home to some of the world's most charismatic wildlife, these icy landscapes have fascinated scientists for at least a hundred years. Buried in snow, ice, rock and oceans are secrets that are being unlocked by scientific investigation. British Antarctic Survey has changed our understanding of Planet Earth."

Those were the opening words to an inspiring video that David and I played many times to many different audiences. It was powerful because it managed to convey something of the awe and wonder of Antarctica, its global importance and its fragility, and the fascinating and impactful science being undertaken there.

Sadly, climate change has come to dominate that science. In 2023, the global warming reached 1.45°C. Weather and climate extremes are now affecting every region of the planet, but the changes in the Polar Regions are particularly stark and profound.

The Antarctic Peninsula has seen numerous extreme heatwaves in recent years, including February 2022 when a temperature of 13.7°C was recorded on King George Island. The following month an unprecedented heatwave struck East Antarctica, with temperatures soaring 39°C above normal. Until recently Antarctic sea ice seemed to be avoiding the dramatic losses seen in the Arctic, but in 2023 its extent plummeted, with devasting implications for the wildlife, from penguin chicks to krill, that depend on it.

The Polar Regions are slumbering giants – they take time to wake, but once risen the scale of global change they can induce is terrifying. David describes in this book his reverence for Antarctica, and the feelings of both privilege and responsibility shared by many scientists who dedicate their lives to studying the remote, frozen continent.

His particular focus was the fate of the ice sheets covering Antarctica and the dire implications for global sea level rise. Mass loss from Greenland and Antarctica is now responsible for more than a quarter of all sea level rise, five times what it was 30 years ago. Ice melt in West Antarctica, largely driven by the warming waters of the Southern Ocean, is thought to be inevitable for rest of the century no matter how much greenhouse gas emissions are cut. But the extent of global sea level rise will depend significantly on exactly what happens there.

David's leadership of major international research programmes such as the EU-funded *Ice to Sea* and the *International Thwaites Glacier Collaboration* significantly advanced our understanding of this and has left an ongoing legacy of important scientific discovery. The work is of global significance because hundreds of millions of people live in at-risk coastal megacities such as London, New York, Mumbai and Shanghai and more than a third of the global population live within 100 km of the coast.

He writes that he hopes over the years he has "been effective in persuading just a few souls that climate change is a real and present danger". He did very, very much more than that. He was one of the leading British contributors to the Intergovernmental Panel on Climate Change and he worked tirelessly to go beyond the scientific community to engage and influence the general public, businesses, and policymakers. His message – as is clear from this book – is so compelling precisely because it comes with the honest authenticity of a scientist who literally bore witness to the changes of global significance that few others have directly observed.

Professor Emily Shuckburgh OBE FRMetS

Director of Cambridge Zero, the University of Cambridge's major climate change initiative

Professor of Environmental Data Science at the Department of Computer Science and Technology

Fellow of British Antarctic Survey.

Prologue

This book tells a small tale. I wish it were grander, but it is the one I have got. So, we, author, and reader, are stuck with it. It was written quickly and, at least prior to editing, was of grossly uneven quality, significance, and coherence.

For reasons that will become apparent, I have lent heavily on readers and editors, and acknowledge their work and insight improving massively all aspects of the work.

"How complicated and unpredictable the machinery of life really is."
—**Kurt Vonnegut Jr Cat's Cradle**

Since I began talking to people about climate, I have discovered a truth that will be entirely self-evident to psychologists, negotiators, and counsellors. A scientist who presents evidence and fact without advocacy is more likely to convince people than one who is steering the listener to their way of thinking as to the solutions. Credibility is eroded where scientists are too fond of their own hyperbole or are obviously selling a political solution – and when it comes to climate change, all solutions are basically political.

However, many times I have worried that this position of avoiding advocacy to maintain credibility can be argued to be at best lazy and at worst a cop-out. It is defensible but lacking in grit and many of my scientific colleagues believe so. It has certainly been a convenient position for me to hold, as it allows me to avoid making decisions about what I believe is the best route forward in dealing with all the big issues that face us. Climate change is only one of the great issues concerning economic development, geopolitical power struggles, and even human motivation. There is little point in us developing ways to deliver plentiful sustainable energy if nations use it to wage wars that turn the Earth and our lives to ash by a different means, or if we use emissions-control measures to bind generations in developing economies to living standards that we in the West have not tolerated for a century.

Remaining publicly neutral about the solutions has served me and, perhaps, my audiences well. Indeed, I hope, over the years I have been effective in persuading just a few souls that climate change is a real and present danger. However, I have often been asked after presenting my facts, what would I do about climate change? What do I do? I will answer, but always prefaced by a caveat that these are personal rather than scientific responses. There are, after all no 'correct' responses.

It is very clear that there is no single approach, and certainly no single solution to climate change, that will not negatively impact a major proportion of the world's population for many generations, lowering living standards, opportunity, and equity across the poorer sectors of our global community. Democracy is a uniquely poor instrument to deliver policies whose near-future impacts are unpalatable, let alone painful. So, while democracy remains effective only at the scale of the nation state, it is nearly impossible to imagine one nation voting to share their advantage with another.

Several recent books make the apparently revolutionary case that we persistently measure and describe the economic health of societies and nations in brute terms. We use metrics that take little account of real lives. These metrics, including gross domestic product (GDP), signally fail to differentiate between societies that stress their populations into an endless and growing cycle of earn and spend, and those with values that promote contentment and happiness. Reading these well-intentioned and truly sensible accounts, what surprises me is not the revolutionary nature of their argument that we should move our discussion away from simple metrics and replace them with ones that express quality-of-life variables – surely a no-brainer – but that we did not do this many decades ago. Rather we remain in our dim dwellings obsessed by numbers that mean so little, except how much the latest iPhone model will set us back.

However, the fact that these measures have such persistence is no mere happenstance. Sadly, their short-term and trivial value reflects many people's trained obsession: the latest fashion, a faster car, a bigger TV.

Perhaps the biggest challenge facing climate change activists is to change the minds of a populous, seeding in their minds better ways to judge the value and quality of their lives, ones that emphasise community, meaning and purpose, and play down wealth, status, and celebrity. Such resetting of a nation's values system is fundamental if we are to hope that people will act altruistically, and within a democratic system. That is precisely what we are asking them to do in facing the climate emergency effectively. Afterall, without a religious framework, we are all simply individuals acting in our own self-interest.

Given our recent observations of major physical and ecological change around the world, it is hard to dispute that we have done both ourselves and our planet some major damage.

A while back, I drew up a dual timeline of the history of Antarctic exploration and the pollution recorded in the ice. It shows that even before the Antarctic continent was first sighted in 1820, CO_2 levels in Antarctic snows were rising above their natural range. In addition, and more critically, the carbon-13/carbon-12 ratio in CO_2 in snow, changed from its stable natural benchmark. This change is the smoking gun that incontrovertibly links the burning of fossil fuels with the rise in CO_2.

Together with long-term measurements, environmental changes including those that I reported in Antarctica in the 1980s and 1990s and the many that have come since, give us a sound understanding of how greenhouse gas concentrations control climate on our planet. We humans have dramatically altered those concentrations.

In a global fashion, and over a remarkably short timescale, we have permanently achieved a comprehensive and impressive buggeration of just about everything that keeps the planet's human, animal and plant inhabitants thriving, healthy and safe. Well done us. We have surely proved ourselves to be masters of the universe, trading our birth right for life-improving luxuries such as carpet fresheners or fizzy drinks, or annual upgrades to our mobile phones.

What still surprises me is the degree to which blame rests with such a tiny fraction of the world's population. It is mostly the citizens of industrialised countries, including our parents and grandparents back only a handful of generations, who have unwittingly fucked our once beautiful planet.

A report by Oxfam and the Stockholm Environment Institute estimated that between 1990 and 2015, the richest 10% of the global population (around 630 million people) were responsible for about 52% of global emissions. Even more shocking, the wealthiest 1% of the world's population, just 63 million people (a number equal to the population of the UK) were responsible for the emission of more than twice as much CO_2 as the poorer half of the world[1] (around 4 billion). For reference, the richest 10% are those with incomes above about £27,000 a year, which is well below the median wages in the UK; the richest 1% are people earning more than about £77,000 per year. By this definition, I was in the richest 1% for eight years of my working life, and thus I accept my oversized share of the blame.

Even the most remote snows on which I walked in Antarctica contain a measurable and permanent record of the lead we added to petrol, the radioactive fall-out from the bombs we tested on Pacific Islands, the

---

[1] - https://www.theguardian.com/environment/2020/sep/21/worlds-richest-1-cause-double-co2-emissions-of-poorest-50-says-oxfam

insecticides that washed off our lands and into the sea, the degraded plastic from our clothes and tyres and, most significant of all, the CO_2 and methane we have emitted from our cars, power stations, factories and farms[2].

We have it in our power to limit the damage. So far there is little evidence that our society, either collectively or individually, is ready to give up very much to achieve this. Perhaps, this is a measure of our own greed, but perhaps it is a failure of society itself. As Jean-Claude Juncker, former president of the *European Commission* wrote in *The Economist* (2007 The Quest for Prosperity), "We all know what to do, but we don't know how to get re-elected once we have done it."

So, what to do? Having spent my entire career committed to providing information without campaigning for a particular policy-response while remaining largely silent on the matter of what we should do about climate change, I have perhaps missed my chance. Given my medical prognosis, and the limitations of my energy, it is too late for me to take any significant role in protest, campaigning or lobbying for investment in green energy or emissions reduction, as I had once planned. Indeed, there are sufficient voices calling for these things at a collective volume that none can claim to have misheard on the central issue, but there are some areas where the science community, in which I have lived and worked, might usefully focus and re-focus in coming decades.

It is vital that science continues to be the foundation for rational debate, and to do this, scientists must maintain and guard their credibility. We will lose that credibility if we use our positions as scientists to add weight to individual policies. Worse still, supporting unjustifiably optimistic views will also erode that credibility. Failing to acknowledge that while there is still much to play for, time has run out and bad things are now inevitable. This failure is colluding with governments' intransigence. For example, it allows them to return to the COP climate talks every year and say the time for action is now without actually committing to any of what is required. No scientist wants to be a doom-monger, but credibility is won through honesty. Modifying our message to avoid scaring the punters will, in time, backfire.

Science must be realistic about the questions we can and cannot answer. Science is not a crystal ball. Our models are built and verified using contemporary observations and occasionally geological analogues. Once our planet has begun to change significantly our models will cease to have any authority. There are limits to how far into the future we can rely on projections. And we need to find better ways to express what we mean by certainty and uncertainty.

---

[2] - https://www.rmg.co.uk/stories/our-ocean-our-planet/ice-cores-climate-change .

Scientists, including social scientists, need to develop different ways to communicate and engage with citizens if the world is to properly acknowledge, respond to, and live with climate change. A powerful approach to building a coherent response is to begin with people's own lived-experience and heritage. Acknowledging a national and personal share of the blame and accepting that some of us have lived beyond our means, may lead to some conspicuous resetting of our expectations.

Adopting better ways to quantify the quality-of-life people experience will assist us in re-shaping our societies to be greener and ultimately happier. Society should not and cannot be judged by GDP alone. Faced with a choice of what we are prepared to give up for the sake of the planet we must dig deep and understand ourselves better than we do. Make no mistake, while it is vital, we cannot save our planet simply by investing in green energy; we must accept that our society cannot manufacture its way out of climate change, and business-as-usual will ultimately mean many sectors face 'no business' and 'no future'.

This is not intended as rhetoric, simply the unvarnished and miserable facts as I see them after almost 40 years on the margins of the game. If we choose to do nothing, or simply cannot agree on what to do, our planet will become incapable of supporting our ever-growing population, and until populations and lifestyles have been readjusted, we will be in dire straits.

I am not suggesting that climate change will kill species directly. That will flow from our own greed and selfishness. Without sound agreements to act on civilian systems for the production and distribution of food, energy, and essential products, these systems will break down and fail – as they have visibly done in Europe since the Russian invasion of Ukraine. Military systems of war and genocide will be deployed in the name of peacekeeping. In the end decades of this century more casualties may occur by our own hand, rather than by the direct result of climate change. As in my own body, it will not be the cancer itself that kills me, but the failure of my own internal organs as the cancer makes it impossible for those organs to do their job.

In 2013, I was asked to write an optimistic paragraph for '*Global Chorus*', a remarkable, illustrated collection of 365 daily meditations on some very large and increasingly crucial themes. Editor Todd Maclean wanted the map showing his contributors to include Antarctica, hence my involvement. I was flattered and proud to be part of this wonderful and uplifting project and wrote:

> *"Quite soon those changes will be evident to all, and eventually many parts of the planet will change in ways we can only begin to predict. But in a strange and wonderful universe, the most*

boundless thing we have yet observed is the scope of the human mind, the strongest is the human spirit, and the most hopeful, the sound of our children learning what we do not yet know.

I am optimistic that as a species we will eventually find ways to repair our damaged planet and build ourselves a truly sustainable future. But building that future will take time and many brave choices: if we are not brave enough to begin, let's raise children who are."

I stand by these sentiments, but as every year goes by, I have felt my optimism diminish. I still believe in the inherent goodness of individuals; it is just a shame that the behaviour of society does not seem to add up to the sum of its parts.

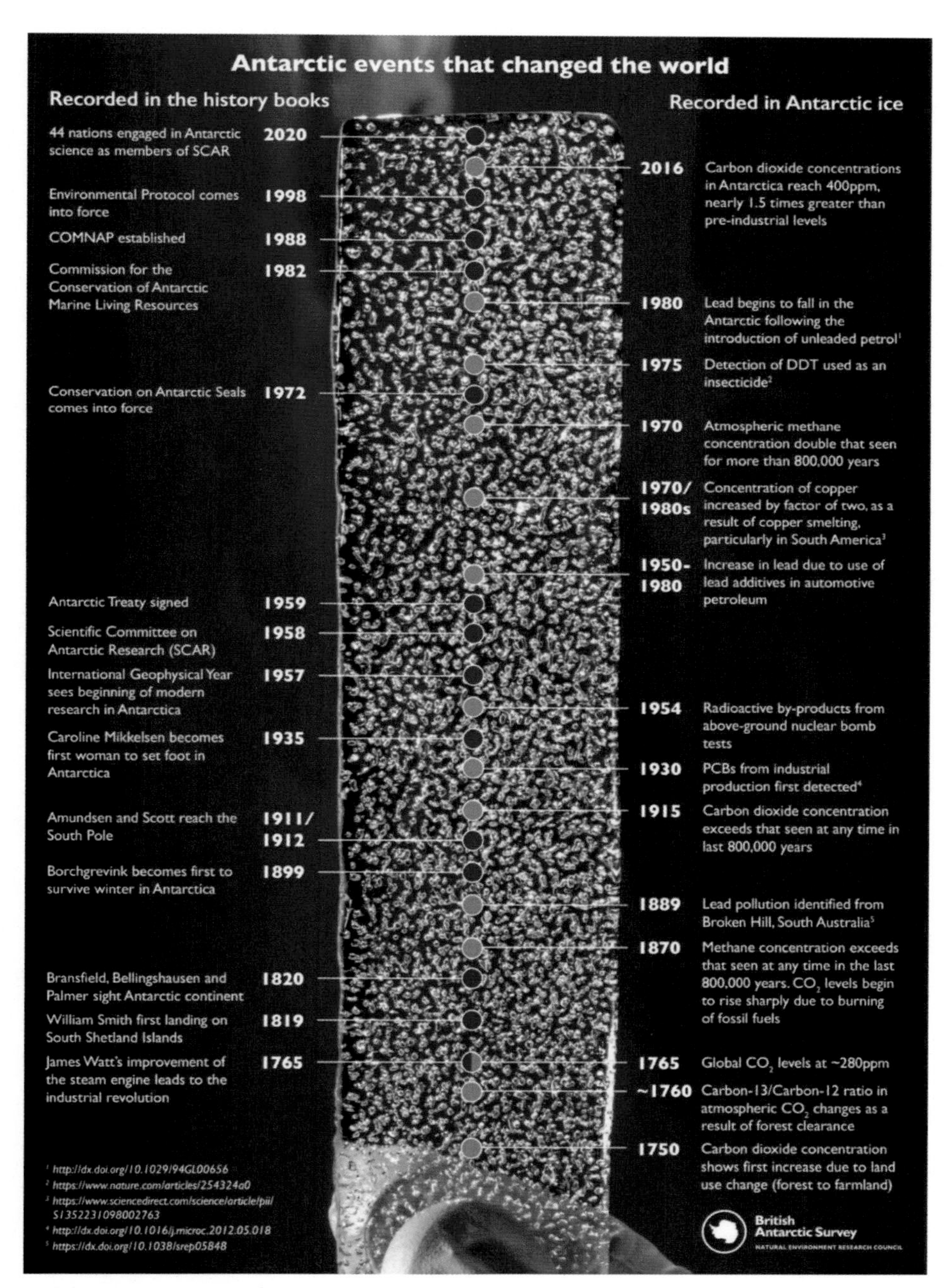

Timeline showing data from history and data captured within the ice core. (Credit: BAS)

PART 1
ANTARCTICA

Camp on the Rutford Ice Stream, January 1986. Behind is Mount Tyree, and just off-picture to the left is Vinson Massif, the highest peak in Antarctica (Credit: David Vaughan/BAS)

CHAPTER ONE

British Antarctic Survey

At the age of 14 one of my career choices was to be a physicist for the British Antarctic Survey (BAS). A few years later at university after reading BAS's annual reports and recruitment information I was intrigued that most posts involved a 30-month deployment to Antarctica.

BAS was only a mile from Churchill College, my University of Cambridge college. I got on my bike and cycled to BAS to meet Dr Chris Doake in the glacier geophysics department. Tall, swarthy, and gaunt, with a permanent half smile, I imagined that Chris spent his youth among the Scottish hills, striding through the heather in summer, and freezing in snow-filled couloirs in winter. He still had a hint of Scots in his accent, and we talked for an hour about BAS, Antarctica and glaciers. By the end I was hooked.

I took a summer job with Chris doing his odd jobs. I read *The Physics of Glaciers*, by W.S.B. Paterson (the only book I could find on glacier geophysics) and applied for a three-year contract in Chris's small group. I was interviewed, did not get the job, but my intention was set. Convinced that I needed to convert from physics to something I could understand, I accepted the offer of a place to do a one-year Masters in geophysics at Durham University. I was one of the last to receive a proper maintenance grant from the government, and at a little over £1400 per year I had just about enough to live on. Supplemented by the wage I earned during the holidays as a sailing instructor, I was able to live comfortably in university halls, although there were few holidays or meals out. I left university without debt, and it makes me sad that today, even the best of our young people on the most challenging of university courses whose value to the nation is significant leave university with a lifetime of debt hanging over them.

In 1985, I applied again to join BAS and was offered a three-year contract as a Glacier Geophysicist. In August the same year I returned to Cambridge to begin preparations to go to the Antarctic that autumn.

I learned to use the equipment we would be using in the Antarctic, including the complicated theodolite. I did crevasse rescue training, dangling on a rope off a cliff in Derbyshire, and attended a truly gruesome medical course, which included stitching wounds, temporary setting of fractures, and some horrific lectures on burns management.

The author practices with jumars, during BAS rope training beneath Curbar Edge, Peak District, summer 1985. (Photographer unknown)

I enjoyed being in Cambridge and being no longer tied to the University. The previous year my friend Chris German had begun a PhD in the Department of Earth Sciences and was supplementing his grant working as a barman in the Eagle public house. One hot evening after his shift we sat in the Eagle's courtyard. Chris is one of the most empathic people I have ever met, and his ability to see into people was scary. He smiled at me as he questioned my motivation for going to the Antarctic. Only half teasing, he told me, "You'll be completely out of your depth at BAS. You're going to be surrounded by serious people who are completely into being in explorers – you're only doing it so when you come home, you'll get laid." If that hadn't been so close to the mark, I might have forgotten it, but it still stings.

He was right to call me out. Even today, I am unclear of what motivated me in those early days. I did not much like England in the 1980s. It was blinkered, dreary and violent. Indeed, from today's perspective it is hard to remember quite how grim things were before the sham optimism of the 1990s eventually took hold. Inflation had peaked in 1975 at almost 25%, and again in May 1980 at 17.8%, and for most of the 1980s unemployment exceeded three million people. Dissatisfaction and resentment were rife, and people seemed increasingly angry.

Most memoirs invoke a level of pre-destiny, vocation, or simple intent that I truly would not claim. Paths emerged and I took them, but the opportunity to escape England, even for a while was welcome. The thing that drew me to Antarctica was, more than anything else, a sense of getting away. I never planned to continue working at BAS for more than a few years, but at the time it suited me, and saved me from working for the oil prospecting company that was the only other offer on the table.

Chris German was even closer in his analysis than he expected. A week later, on a warm summer night, I met him at closing time at the Eagle and we moved on to a house party that someone had told him about – Chris was always getting invited to parties. Through the cigarette smoke and dodgy music, I met a girl. She was pretty, and I was taken by her dark eyes and voluminous bob, but even more so by her quiet self-assurance. Just a little aloof, but completely comfortable in herself she seemed to take a shine to me and with few preliminaries we walked across town to her room down a quiet side street off the main drag.

We talked for a couple of hours. I remember, she was proud of a grand or great-grand father who had been a famous scientist. She was surprised I had not heard of him. We worked out our couple of mutual friends. Then said she had a lecture in the morning and needed to sleep, and she took me to her bed. We parted in the morning, her to her lectures, me to Antarctica. Sometimes to this day I think of her and the closeness and warmth of that one night. If I had been staying around, I would have pursued her, but with my departure imminent, I simply left it at that. Hardly a regret, but certainly a what if?

Two days later, I left for Antarctica and loved it, but Chris was right, I never truly felt I could relax in the company of people whose capability and commitment was so superior to my own. For 35 years, I have always had something to prove and perhaps that is no bad thing.

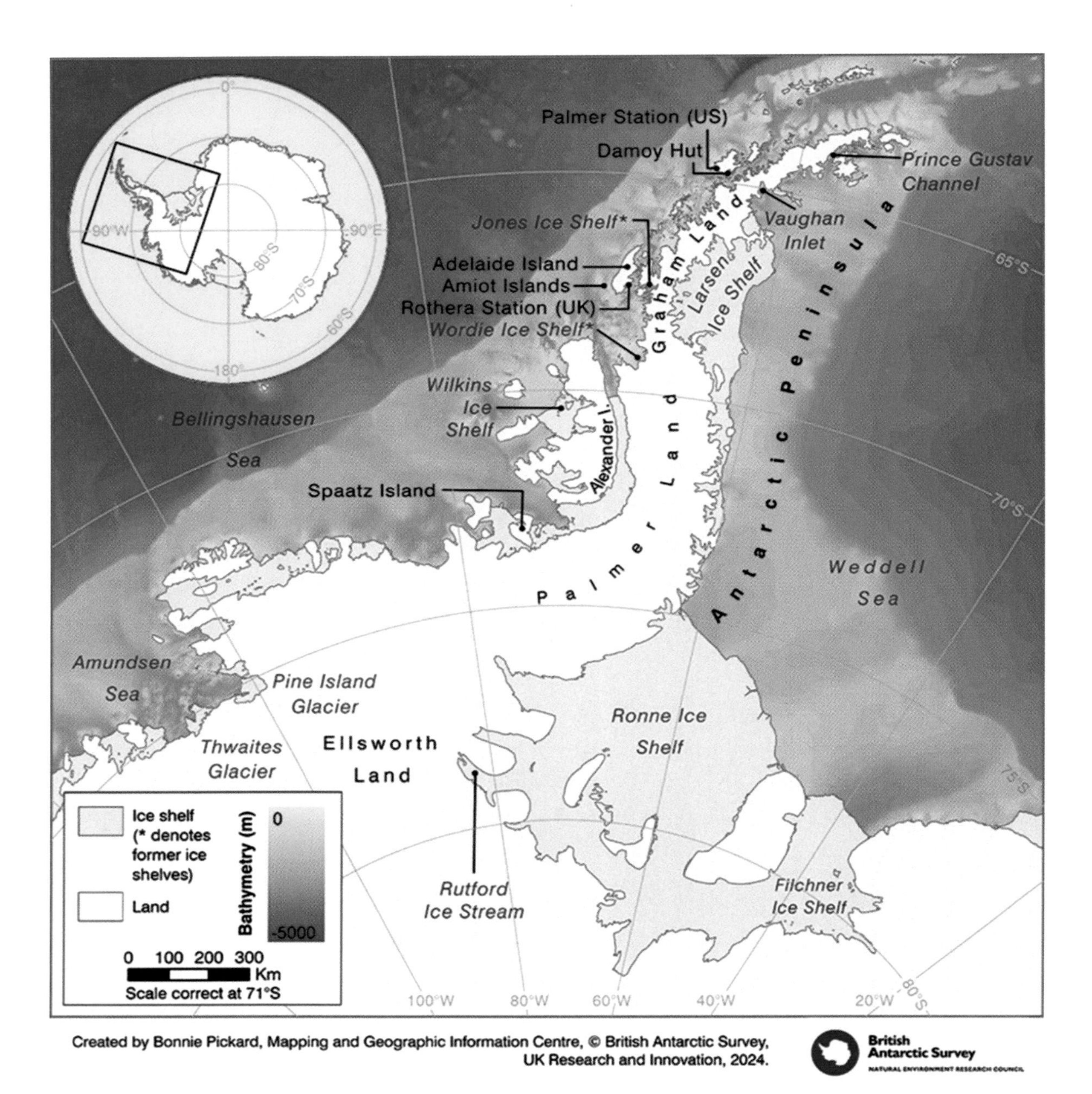

Topographic data from the SCAR Antarctic Digital Database, 2024. Bathymetry taken from GEBCO Compilation Group (2023) GEBCO 2023 Grid (doi:10.5285/f98b053b-0cbc-6c23-e053-6c86abc0af7b). (Credit: BAS)

CHAPTER TWO

Going South

I was one of the first British Antarctic Survey scientists who did not travel all the way to Antarctica by ship at least once. Since the 1940s, all rookies joined the Royal Research Ship (RRS) that would conduct scientific research on behalf of the UK government for the six-week journey south. I was spared the trip, in part because BAS was beginning to question the prevailing notion that a scientist's time was a commodity with negligible value. Indeed, my boss Chris Doake viewed sailing south as a monumental waste of time, although his view was probably coloured (green) by the fact that he was a renowned puker.

In October 1985 I traded an English autumn for the warm spring of Montevideo, Uruguay. I arrived with six others on a commercial flight from London Heathrow. Montevideo had the air of a city several decades older than it was, post-colonial and showing signs of both beauty and decay. Vintage American cars and trucks filled the streets, and occasional groups gathered for their own pleasure on the street corners to play music and dance the tango.

There on a hot and sweaty afternoon six of us roamed the streets and ended up sitting in the Plaza Independencia under a huge portrait of Che Guevara. A man in a white suit and a Panama hat asked if he could take a photograph of us. He had a huge box camera on a sturdy tripod and with a flourish he exposed a glass plate using a button on the end of a long cable. Then he turned the camera round and seemingly inside-out. Within minutes, he had developed the film and printed multiple copies of the image inside the same box. He sold us the prints for a couple of dollars each and I sent mine to my mother. It was the last she would see of me until the following April. She kept it among her treasured things and when she died thirty years later I found it in a box. It is faded and creased, but the image is still sharp: six of us standing scruffy, happy, and very young.

Plaza Independencia, Montevideo, October 1985. (Credit: David Vaughan)

We stayed two nights in Monte waiting for the RRS *John Biscoe* to arrive from the UK. The first night we ate in a restaurant in a local meat market and drank cerveza. My friends and colleagues ordered steak. This seemed to me too easy and seeking the full South American experience I ordered randomly from the menu. For my boldness I was presented with a Chilean delicacy called chinchulín, a type of chitterling sausage made from sections of roasted intestines stuffed with offal. The chinchulín that lay before me was grey, gritty, and frankly disgusting. I am sure that they are beloved by a few, but I was roundly mocked by my steak-eating comrades. In truth, I struggled to finish chinchulín, spurred on only by my instinct to save face. I never tasted Uruguayan steak, but I am told it is the best.

On the second day, the RRS *John Biscoe* and RRS *Bransfield* arrived. During the evening we joined a few of the ships' crew and toured some very different bars. If you know where to look, such bars are found in every port, servicing the needs of sailors. Outside one hung a distinctive sign consisting of a dead penguin hanging from a hook. The carcass dripped goo onto the pavement below. It smelled bad, really bad. The bar was apparently a favourite of those travelling to Antarctica. We went to many similar bars, all of which seemed to have young waitresses who seemed more interested in getting to know the clientele than in serving drinks.

Several of our party were due to stay in Antarctica without female company for a 'full tour' of 30 months. For some, temptation was strong. Several deals were done, and various lads skulked off. For at least one of them, the evening ended in the classic style: a big bloke emerging from a wardrobe and demanding the seaman's wallet. I stuck to cerveza and local whisky. Far too much. Most of it lost in the street behind one of the bars. As dawn arrived, I crawled back to the ship and up the gang plank.

I should point out here that when I first joined BAS, jobs involving Antarctic work were offered only to unmarried men between the ages of 21 and 35. Men held all the key leadership positions at BAS HQ where women were employed almost exclusively in secretarial or administrative roles. By modern standards Antarctic stations and ships were all-male, rough, and rather boorish places.

Thankfully, things began to change in 1983 when lead cartographer Janet Thomson became the first female scientist to work with BAS inside the Antarctic Circle. By 1986, Professor Elizabeth Morris was Head of BAS's Ice and Climate Division. At Signy Station, on the South Orkney Islands a female-only team of scientists and support staff completed a summer field season. At the beginning of the new decade women were working deep field. In 1993, women overwintered at Signy Research Station, and by 1997 the first women overwintered at Rothera Research Station. In 2013 Professor Jane Francis broke the glass ceiling to become Director of BAS.

Today, women, ethnic minorities, gay, trans and cross-dressing people enrich the BAS community, making it kinder, more supportive, and much more fun than it was in those early days.

Meanwhile, back to 1985 and Montevideo. As we readied to set sail for Antarctica one of our number, who was nicknamed 'Metal Micky' for his prominent gold tooth, woke late after our night out. Disorientated and monstrously hung over he was in the windowless library below decks on the RRS *John Biscoe*. He was, however, on the wrong ship! By the time he emerged the RRS *Bransfield* had dropped its mooring lines, raised its gang plank and was heading out into the mouth of the River Plate. Once the mistake was exposed, the ship waited on a mooring while a launch was procured to take Metal on his trip of shame out to meet it. The ship's mate delivered him to the Master for a severe telling off. I am sure that this report remains on Metal's file somewhere at BAS HQ.

On the RRS *John Biscoe*, and with hangovers intact, we took to sea and headed south along the coast of Uruguay and on past Argentina. RRS *John Biscoe* was a small ship by modern standards. In 1956 it replaced its predecessor, which bore the same name. *Biscoe* was used by BAS to move staff, resupply its stations and occasionally as a scientific platform for marine biological studies. While small and somewhat underpowered she was compact, homely, and a favourite of BAS crews.

The complement of scientists and support staff enjoyed relative freedom, providing we steered clear of the crew's bar and the officers' mess. We had our own mess room, which in the evening became the bar. We slept four to a cabin, which was piled high with our huge kitbags. On deck, drums of petrol and Avtur (aircraft fuel) were lashed down on every spare metre of deck, to the chagrin of the smokers who were banned from smoking on deck.

As the days passed, we became accustomed to the ship's routine, whose metronome was the galley serving three good meals each day. We stood for hours on the 'monkey island', the viewing platform above the ship's bridge. There, we were out of the way of the officers, but could see everything around us and everything happening on deck. We spent many hours watching and identifying the birds, and with each day we progressed farther south the temperature fell.

Small ice floes viewed from the monkey island on RRS *John Biscoe*. (Credit: David Vaughan/BAS)

Saturday night 'sundowners' aboard RRS *John Biscoe*, on first sight of the Antarctic Peninsula. In 1985, packing instructions issued by BAS included instruction to include a 'tropical weight sports jacket', in the event any official duties were required to be undertaken. This fine example from the Oxfam shop in Regent Street Cambridge accompanied me on many trips but was never called on for 'official duties'. (Credit: David Vaughan/BAS)

Below decks, we sprawled lethargically in the bar. I played a lot of Scrabble. We read and listened to music. Few of us had headphones, so the deal was struck; once someone slid a cassette into the stereo, it was played to its end or until that person left the bar. A certain amount of grumbling and muttering was allowed, but once it started it there would be no stopping, no matter how awful the choice. It was a tradition which widened my appreciation of some genres and deepened my loathing of others.

One day I was beckoned by one of the ship's crew into his cabin, an able seaman called George. Seamen and scientists were forbidden to have alcohol in cabins, but George did not like to drink on his own, and being on a night shift, out of synch with the bar hours and his crewmates, George was not fussy about who he drank with. He was a broad Glaswegian, another to call me 'laddie', and as small talk seemed superfluous, we sat and drank in silence for a while. Eventually he asked, "what did you do before this?" Happy to find some topic we could talk about; I replied that I had been in Durham for a year to do a Masters. Without irony, he admitted that he too had been in Durham to do "six months for grievous bodily harm".

In late October, I spent my 23rd birthday crossing Drake Passage in a storm. It was notable only for upending a large mug of tea into my lap as the ship lurched on a wave. We made landfall in Antarctica at the end of the month, and 14 of us disembarked to Damoy Hut, in Dorian Bay on the tiny Wiencke Island. Here we would wait for an aircraft that would pick us up and fly us south, over the lingering pack ice to Rothera Station.

Rough weather aboard RRS *John Biscoe* in the Drake Passage, October 1985. Shortly after this photo was taken an officer came down from the bridge and told the author in certain terms, not to be an idiot and to go inside. (Credit: David Vaughan/BAS)

Damoy Hut had only two rooms, one for living and one in which 14 of us would sleep in three-tier bunks fabricated from old packing cases. Nearby stood a tin hut listed by Argentina as an emergency refuge. It was in such state of disrepair that any refuge taken therein would be truly desperate, so we fixed the roof and installed our toilet buckets in it.

Within 24 hours, we settled into a routine of passing weather reports to Rothera and waiting for the aircraft to come and collect us. We completed the annual census of the local gentoo penguin colony. In the name of training we built an igloo and slept in it, did some practice ropework and crevasse rescue. Crevasses are rifts in the ice produced by stresses built up as a glacier moves. Small ones might be a metre wide and a few metres deep, but a large crevasse can be 50 metres wide and 50 deep, deadly and an ever-present concern in Antarctica. Crevasses are most dangerous when hidden by 'snow-bridges', which are hard to spot in less than perfect weather conditions potentially too weak to bear the weight of a person or vehicle, so crevasse rescues are vital training when working in Antarctica.

I tried skiing for the first time. We used an ancient skidoo (motorised toboggan) as a ski lift and although snow conditions were rather icy most of us managed a couple of runs. Almost instantly I fell on the hard, icy snow and dislocated my shoulder. I immediately knew I had done something serious, so attempted the manoeuvre that the doctors had described during our medical course in Cambridge. Although it hurt a great deal, it achieved absolutely nothing and seeing that I was in trouble, my colleagues wrapped me in blankets and returned me to the hut. The doctor at Rothera was contacted by radio. He said that given the lack of trained personnel available, rather little could be done for me. I was sent to my bunk with a couple of codeine and after 12 hours of troubled rest, the shoulder relocated itself with a clunk. It was stiff and painful for several weeks, but it was to my advantage that the injury was not seen by the doctor. He might well have taken a dim view of my onward deployment into the field and called for my repatriation. As it was, I soldiered on with paracetamol. Since then, I have suffered many recurrences in both shoulders, several in difficult circumstances.

After waiting for 17 days, sending Rothera hourly weather reports by radio, there was no suitable weather window to allow our uplift by aircraft and it was decided that the *Biscoe* would return to pick us up.

Once again we were onboard ship, but it soon became clear that the sea ice conditions were not improving as rapidly as hoped. The so-called inside route south to Rothera was impassable, so our Master, Chris Elliot, decided to take the longer route to the west of Adelaide Island.

In the days that followed we made several attempts to push through the ice around the southern tip of Adelaide Island, only to be held up by the thickness of the sea ice. Eventually, we ground to a halt and our little ship was beset in ice.

Icebreaking ships, even small ones like the *Biscoe,* work most effectively when they can take a run up at an ice floe, gain a little speed and ride up onto the floe before bearing down to split it. They then push forward to open the crack, back up and repeat the process. The ice that gripped us was not typical for the region and time of year and was not so easy to dispatch. It consisted for the most part of a broken rubble of ice, which quickly flowed like porridge into any open space formed around the ship. Full astern or ahead produced only a few yards of movement before the icy porridge pushed into a ridge barring further progress.

Several days passed and we remained in the firm grip of the ice. We watched and waited. Up on the monkey island a continuous debate ran on the properties of the sea ice, our plight, and every possible outcome.

Eventually the US Coastguard's new ship the *Polar Duke* was called to our assistance. No bigger than the *Biscoe*, but with considerably more power she could just make way through the ice to join us.

Many attempts were made to clear a way to turn the *Biscoe* so that we might follow her wake out of the ice but without success. The sea ice simply flowed back into any space that the *Polar Duke* managed to open. Eventually, with the sea ice and both ships drifting towards the small archipelago of the Amiot Islands, a steel hawser about the width of my arm was sent between the ships, and an attempt was made to tow us out.

The *Polar Duke* applied pressure; the hawser took the strain. A second before it parted, the guy standing next to me said quite plainly. "Cables that long and heavy shouldn't go that straight". The cable broke with a canon shot and recoiled faster than any eye could see. The ragged ends sprang back along its length towards and over each ship, leaving a perfect vapour trail, like that from a jet engine hanging in the air. My friend George, who was on the foredeck as the cable passed him was very lucky – unscathed but visibly shocked.

This was a last and desperate measure, and minutes later our Master took what must have been the awful and momentous decision to abandon ship. We were gathered and told to pack a single bag for transfer to the *Polar Duke*. Later, it was telling to discover what everyone packed. Warm clothes obviously, but some items of sentimental value. At least one scientist packed his collection of Leonard Cohen tapes, another left the only draft of his thesis behind on the *Biscoe*.

Some wooden pallets were brought up from the hold and lowered to the ice to form a bridge. I was one of a small team sent onto the ice to carry the bags across to the *Polar Duke*. This was soon accomplished, and then one-by-one the ship's scientists, crew, and officers clambered across the ice and up the ladder to the safety of the US ship. We stood in silence at the rail of the *Polar Duke*, watching as the Master left his ship.

US research vessel, *Polar Duke* attempts to make a path for the beset RRS *John Biscoe*. Close to Amiot Islands, Antarctic Peninsula (December 1985. Credit: David Vaughan/BAS)

In that instant it seems that the very nature of the ship changed. From a living entity with a personality and even a gender, it became a dead, or at least dying, hunk of metal. And we were sure that this was the last anyone would see of it.

The *Polar Duke* turned for the open sea and crawled out of the ice, leaving the *Biscoe* abandoned in the ice. Its anchors were left hanging in a forlorn hope that they would snag on shallow water before the ship itself went aground, but with little hope that the anchors would hold against the weight of ice.

Abandon ship! Bags being transferred from RRS *John Biscoe* (left) over sea ice to *Polar Duke* (right). The author is on the ice, left of centre (December 1985. Credit: Rick Frolich/BAS)

At home in South Devon, my parents read in *The Daily Telegraph*, that the BAS had abandoned ship in Antarctica. To save my mother's concern, my father lied and said I was on a different ship, until he could sneak off and get on the phone to BAS HQ to confirm her precious boy was indeed safe.

The *Polar Duke* took us a couple of hundred miles north again to the US Antarctic Station Palmer. There we were welcomed warmly and very quickly a party was arranged. A large warehouse was set up for music and dancing and we found other corners of the station to unroll our sleeping bags.

In the early hours, I was standing outside the party when a huge man came out to pee in the snow. I recognised him as the chef from the *Polar Duke*. He was too far gone to talk. He began to do his business, but then simply fell over and did not get up. His immense size and level of intoxication meant that I could do nothing for him on my own, so I went for help and returned with a couple of others. We managed to move the now properly unconscious form onto a trolley and rolled him down to the wharf, but the gang plank was far too narrow for us to get him onto the ship. We hailed the night watch, who without hesitation jumped into one of the deck cranes, lifted a cargo net out of the hold and dropped it next to the inert body.

The last I saw of the cook was an arm poking through the net as it disappeared into the inky darkness of the cargo hold. No more was said, but I heard that breakfast was served as usual the next morning.

After a few days the *Biscoe* was located by an aircraft pilot in thinner ice and retrieved some days later by a skeleton crew taken in by the massive German icebreaker, the RV *Polarstern*, who led her out of the ice. The *Biscoe* continued in Antarctic service with BAS until it was replaced in 1991 by the RRS *James Clark Ross*.

Rothera Research Station as it was around the time of my first arrival there. (Credit: Andy Smith/BAS)

CHAPTER THREE

Preparing for fieldwork, Rothera Research Station

I arrived at Rothera Research Station on a BAS Twin Otter aircraft after a flight from the US Palmer station. In those days, Rothera was a ramshackle collection of wooden buildings and huts on a rocky promontory on the otherwise snow-covered Adelaide Island. Home to around ten overwintering staff comprising chef, doctor, carpenter, plumber, radio operator, electrician, mechanics, and a few field assistants, or 'polar guides' as they preferred to be called. During summer these numbers swelled to around 50 with new winterers, scientists, pilots, aircraft mechanics and operations planners.

Conditions were crowded and basic if not a little primitive. The water we drank and washed in was provided from a big tank of melted snow which had to be topped up several times each day with shovels. Our toilets were buckets which needed to be emptied by hand, a job that was shared by everyone on station.

A tradition at all BAS stations and ships was to nominate a 'gashman' who, in addition to their normal duties, would, on a specified day, pick up domestic tasks, including washing up, washing floors, clearing debris from the bar, emptying bins, assisting the chef and, most dreaded, emptying toilet buckets. Even BAS Directors were not exempt.

Each day full buckets were hauled across the gravel beach and out onto the ice, before being emptied through cracks in the ice into the sea. If any trace of the contents remained it would attract birds, especially sheathbills and skuas. The birds would eat it and likely paddle it back into our supply of fresh snow for the melt tank, with potentially drastic consequences for our health. With water in short supply, we were limited to one short shower each week, and clothes were washed only when necessary. The station thus had a generally musty air about it.

My main tasks at Rothera were to complete my field training and help prepare the scientific equipment for our upcoming field season. Working through the lists that my colleague Rick Frolich and I had begun in Cambridge, we checked every item of kit, that we had spares and maintenance manuals, and that everything was packed properly for the onward journey south.

The author (left) prepares field equipment in workshop at Rothera Station. December 1985. (Credit: Charles Swithinbank/SPRI)

They were long days, but it was time to go. The skidoos and sledges, camping equipment, scientific boxes and fuel were divided into individual loads each weighing around 2000 pounds. As is common practice around the world, aircraft operations are conducted in a curious mix of units: pounds for weight, nautical miles for distance etc. These loads were transported up a hill to where the aircraft operated out of a primitive snow runway, or 'skiway'. A couple of days later, I clambered into the back of one of BAS's three bright red aircraft and strapped myself into a folding seat among assorted jerry cans and boxes.

BAS acquired three de Havilland Twin Otters in the 1970s and early 1980s, and another in 1988. Despite hard lives involving many incidents, repairs, and modifications, these four aircraft remain the backbone of BAS Antarctic field operations. They have a warm place in the hearts of BAS folk who have flown in them. Their continued durability after forty years of service is testament to these 'airborne Land Rovers'.

The aircraft taxied and heaved itself into the air, climbing into the blue skies above Rothera. For the first time I got a sense of the grandeur of the snowy mountains of the Antarctic Peninsula, stretching to the north and south. We flew south, briefly refuelling at Fossil Bluff, and onward down George VI Sound to Spaatz Island.

One final obstacle remained between our small team and our fieldwork proper. This was the small matter of extracting the aircraft fuel required to deploy us and our equipment to our field site on Rutford Ice Stream.

The previous year, the RRS *Bransfield* had unloaded six hundred 40-gallon drums of aviation fuel each weighing a little over 400 pounds onto the ice at Spaatz Island. By the time we arrived almost a year's snow had fallen, and the drums were well buried. The four of us flew to Spaatz Island to dig out the drums and make them ready for use in our onward deployment. We were told by the Rothera Station Base Commander, that deployment would begin when we got 350 drums to the surface.

Having pitched our tents, we selected our shovels and began digging. It was hard work with nothing but a small sledge, and a skidoo to drag the drums. At 2.5 metres we reached the top of the first drum. A swift calculation suggested that the depot was covered in about 400 tonnes of snow.

Beginning excavations on Spaatz Island, December 1985. The top of the first drum of fuel is visible at the feet of Rick Frolich and the author. The snow above it had fallen in a single year since the depot was established (Credit: David Vaughan/BAS)

Over the coming days, we devised a system that meant we moved only a fraction of the total mass of snow. This involved digging a central trench the width of two drums, and then mining out the two either side without moving the snow above. Moving the drums was accomplished using a small sledge towed by skidoo. It took almost three weeks to bring the required 350 drums to the surface. Any lingering concern over my shoulder dislocation faded with this extreme form of physiotherapy.

Our Spaatz Island bargain with Rothera was complete. Drums of avtur, aircraft grade kerosene, were lined up next to a line of flagged bamboo poles that delineated our makeshift airstrip. An aircraft was sent to take us to our destination, Rutford Ice Stream.

The author with the first polaroid photograph of the radar echo from the base of Rutford Ice Stream. Behind is the first version of the radar reflector, which was later more than doubled in size. (Credit: Rick Frolich/BAS)

CHAPTER FOUR

On the ice!

Leaving the mountains of the Antarctic Peninsula behind, we flew on into West Antarctica, where mountains gave way to the open expanse of the ice sheet, broken only by the occasional 'nunatak' of grey rock.

We landed four hours later. It was Christmas Eve. I unfolded myself and jumped down out of the bright red Twin Otter into soft snow, next to a hastily off-loaded pile of boxes containing food, scientific equipment, generators, and the other assorted paraphernalia of camping life. It was at this point that I finally understood for the first time the seriousness of the undertaking. We pitched our tent rapidly, tested the radio, and then the pilot and his mate jumped back in the plane, which now without cargo and light on fuel, took off in seconds, and within a minute was a rapidly diminishing red dot.

That first season, we were a party of four. Me as the junior scientist; Rick the senior scientist - a rather dark soul and active socialist; Jonathan Walton, a professional surveyor and long-experienced Antarctic hand; and Damo, a taciturn field assistant and safety expert.

Once in the field, we were very much on our own, a self-sufficient team with two living tents, four skidoos and six sledges. BAS had many similar field parties deployed across the Antarctic Peninsula, and throughout Graham Land. To avoid confusion, each had a simple alphabetic designation, Sledge Alpha, Sledge Beta, etc. We were Sledge Romeo, and Sledge Sierra. Rutford Ice Stream was the most remote field site that year, almost 1609 km from Rothera, and around 400 km to the north of the nearest other field party, a geologist, and his field assistant.

Rutford Ice Stream had been the focus of BAS glaciological research for several years. It is one of the major glaciers that drain the West Antarctic Ice Sheet. Our task was to travel up and down its length (150 km) and across its width (35 km) to locate and take measurements at each of the 120 four-metre aluminium poles planted in the ice a year before. From our measurements we would eventually derive how far the poles had moved during the intervening year. The distortion of the network would eventually tell us a great deal about how the glacier was flowing, and how the forces along its sides and base resisted flow. Ice streams and glaciers are basically the same, except that most glaciers have exposed rock on their margins, while ice streams are embedded within the ice sheet.

With three weeks on Spaatz Island under my belt, I was already accustomed to life in a BAS field camp, which was to be my home for the rest of the summer. Our bright orange two-person tents were pitched directly on the snow, and just large enough for two inflatable mattresses, a radio, a stove, a clockwork alarm clock and three boxes containing food, cooking equipment and 'luxuries'.

Although small, they were a tried and trusted design evolved from those devised by Scott. They were far from lightweight, and the pyramidal design was reputed to be able to withstand winds of 100 miles-an-hour. They were held down by eight metre-long stakes planted firmly into the snow and an apron, or valance, onto which would be placed heavy boxes or blocks of snow that would add extra weight and prevent winds getting under the edges. A generous gap between the inner and outer layers of fabric helped with insulation, and a pipe near the apex allowed all types of noxious fumes to escape. The tent's entrance was a tunnel of fabric which was tied closed, a simple solution that avoided the need for a zip or anything whose breakage might cause problems in times of emergency.

I shared a tent with Jonathan. Inside, I quickly became obsessive about keeping the place tidy. Outer clothing and boots were stored between the inner and outer layers of the tent close to the door. Here too, on the left was spare fuel for our primus stove and an old porridge tin in which to pee; on the right we stored blocks of clean snow ready to be melted for water.

While outside the cold is ever present, a pyramid tent is a sanctuary in which all the requirements of life are within arm's length. Stripped to its essentials, it is simple and straightforward. With clothing hanging to dry in the apex of the pyramid, and the primus stove roaring for tea, there's no better place.

In 2000, I gave a talk at the Royal Institution in London, presenting two beaten up brass cooking stoves, of elegant and unarguably old-fashioned design. I challenged my audience to identify the one the I had used six months earlier on my last trip to Antarctica, and one (loaned from the Scott Polar Research Institute) used by Otto Nordenskjöld during a Norwegian expedition to the Antarctic Peninsula in 1901. My point was that BAS continues to use stoves of an almost unchanged design, because they remain without doubt the best available tool for the job. They are simple and reliable, burn paraffin (a fuel unlikely to cause a fire in a tent) efficiently and quietly, and can be repaired with a few spare washers and a couple of simple tools. Today they sit in BAS field tents alongside satellite phones. My moral: take what works and stick with it, until you're sure that the new thing is better!

Outside the tent's door two shovels stood in the snow. One with green tape on its handle for digging snow blocks for water, and the other with red tape for digging toilet holes. A little way from our living tent, a large tarpaulin protected our two skidoos from drifting snow. Our scientific equipment was stored in a line of boxes marked by bamboo poles with flags.

In all, the BAS field unit is a highly evolved, relatively comfortable, and safe way to camp. It takes a couple of hours of hard work to set up, and a couple more to dismantle and pack onto sledges. But with sledges towed by skidoos, we could travel from site to site as the science required and could be independent for several months at a time.

In that 1985/86 season, we camped for a little over 100 days. Each of our skidoos covered around 2000 km over the ice. By the standards of the day, a long but not excessive season.

Over the years I have accumulated the equivalent of a couple of years under canvas in Antarctica, and for the most part, I remember them as enjoyable times. On most days when the weather was good enough, we worked hard and long, often travelling many miles across the ice on skidoos to complete our scientific tasks. Each evening we would return to camp to eat, speak to Rothera on the radio, copy up field notebooks, and attend to our camp duties, before crawling into our sleeping bags.

The structure of our days was preciously guarded under the 24-hour daylight. The sun was noticeably higher at noon, raising the temperature a few degrees. It was still well above the horizon at midnight and even on a dull day it was perfectly possible to read without light. Twenty-four-hour daylight is a blessing in many ways, accidents are less likely, and it is a convenient aid to nocturnal peeing, but after a few days the monotony of constant daylight plays havoc with the human body clock. To prevent descending into a spiral of fractured sleep and work, we agreed that even if we worked late, we would nevertheless begin the day at the same time. And so, every morning our alarm went off, and one of us leaned out of our sleeping bag to make tea followed by porridge. The rest of the day could begin.

We dressed for the weather and the task planned, but usually, in a base layer of cotton, a woollen shirt, and one or more woollen jumpers. Over everything, we wore trousers and smocks, made from cotton Ventile, a material that was highly windproof but still allowed evaporation of perspiration. On our feet we wore soft boots called 'mukluks'. The inners could be removed for drying. Hats and gloves were a personal choice, but we all wore BAS-issue sunglasses all day and every day to protect us from snow blindness, and the headaches that come from being exposed to the Antarctic sun.

The author dressed for outdoor work on Rutford Ice Stream 1986. Windproof jacket made from closely woven cotton Ventile. Sheepskin hat from Montevideo. (Credit: David Vaughan/BAS)

Communications were limited to a daily radio 'sched' in which several field parties were called up each day at the same time by the radio operators at Rothera Station. Due to the often-poor radio conditions, these conversations were normally terse. While we could usually hear the more powerful broadcasts from Rothera, often they had great trouble in hearing our little radio set. We would switch from one frequency to another to find one that was workable. On the worst days, conversations could be brief in the extreme, and Rothera would be satisfied once they could confirm that they knew our location and that we were all well. All our conversations were broadcast on open frequencies, so could be heard by the other field parties, and under exceptional atmospheric conditions by radio hams around the world. And so, we were instructed to always maintain strict radio etiquette; nothing rude, sexual and no mention of alcohol! We amused ourselves with codewords that, we flattered ourselves, were sufficiently obscure not to get us into trouble - 'Tango Sierra' = 'tough shit'. Discussion of our meagre supply of whisky could only be made by reference to 'triangular anti-freeze', a term derived from the shape of the Grants Whisky bottle.

On many days, it was barely possible for us to be heard by Rothera and we would resort to shouting 'Roger, Roger, Roger' in response to their questions. Our contact home was similarly difficult. Until the advent of email in the late-1990s, field parties were allowed a single message of 200-words to and from the UK each month. Outgoing messages were read over the radio and copied down by hand by radio-operators at Rothera. They were then typed up and telexed, later faxed, to BAS HQ in Cambridge where they were torn off and posted to our main contacts. Replies came by the reverse process.

One quickly grew accustomed to the anticipation of the arrival of these messages, and the inevitable disappointment in how very dull their contents generally were. Sadly, the writing of them became something of a chore, as 200 words is too many for a postcard-type message, and too few to say anything significant.

As with all isolated groups, food becomes particularly significant for morale and as a punctuation of the working day. Wooden food boxes about the size of a large shoebox, supplied by a company called Andrew Lusk & Co. Ltd, each contained an identical set of rations for two people for ten days. Dehydrated evening meals of three styles: beef stew, vesta beef curry and mutton stew. We tried to perk up the stews with some basic herbs and spices, but little could cut through the grim gravy that surrounded the reconstituted meat. Each box was said to provide around 5000 calories per person per day, but many of those calories were in the form of excessive amounts of butter and sugar cubes.

Luxuries included were 20 bars of chocolate, tins of sardines and a weird sausage meat. My favourite items were the one tin of bacon in each box, and the fruit biscuits. These were welcome daily treats but needed careful handling as at -30°C, chocolate can reach a similar hardness to tooth enamel. The only item that was never opened was the bottle of diabolical vitamin pills -small black capsules of uncertain efficacy, but very clear consequence. Consumption of even one of these capsules could leave you belching sour cod-liver oil for days.

Having been packed in the UK, shipped through the tropics, and stored several years outside at Rothera Station before use, the food was often not in the best condition. Butter was often sour; sardines were mush, and the chocolate was covered in a powdery white residue. However, we survived, and most of us lost only a few pounds during a field season.

Although it was Christmas Eve when we arrived on Rutford Ice Stream we were far too busy to worry about celebrating. However, a few days later all four of us crammed into one tent, decorated it with what we could find, cooked up the best food we had, opened a couple of presents we had carried from home and opened a bottle of 'triangular antifreeze'.

Christmas 1985 in a pyramid tent on Rutford Ice Stream. Jonathan Walton and the author.
(Credit: David Vaughan/BAS)

Years later I found a solution to the problem of how to ensure that my family received their presents from me on Christmas Day. To avoid the temptation of an early peek, I wrapped presents for everyone and hid them in plain sight around the house before I left for Antarctica. My wife, Jacqui, was given a plain envelope of cryptic clues as to their location. Everyone had a clue or two to work on. It worked well until my father turned solving the most difficult clues into a competitive sport. He began to ransack the house for the unfound presents. Jacqui chased him out of her knicker drawer and absolutely forbade him going into the loft. It was Boxing Day before all the presents were found!

Meanwhile, back on the ice we were in fine Christmas spirits. However, our evening ended abruptly when our resident Socialist Worker Rick took offence at Jonathan's idle wondering as to how much he might earn by buying British Gas shares. We woke the next day to fine weather and finally began our scientific work in earnest.

The author using Zeiss Th2 theodolite 1986. (Credit:David Vaughan/BAS)

CHAPTER FIVE

Freezing fieldwork

Our daily fieldwork was painstaking, very cold and tedious. We made thousands of measurements using our theodolite. Already an ancient instrument, this was Jonathan's pride, and it was treated with the care and reverence normally reserved for a minor deity. I would sit on the back of a skidoo and record in my notebook the angles he called out. Then I had to check that they were correct by adding angles in each triangle within the network formed by the poles and checking that they totalled 180°. I quickly learned to write wearing gloves. After each round of angles, I would jump up and run around in the snow, while Jonathan prepared for the next location.

The author using the Magnavox Geoceiver on Rutford Ice Stream, early 1986. I am operating the instrument inside the box; its antenna is on the tripod next to me. (Credit: David Vaughan/BAS)

Each night, in the comfort of our tent, a second 'fair copy', averaging each set of numbers, was recorded in an abstract book. We also measured distances with a 100-m steel tape measure, and positions with a Magnavox Geoceiver. This last instrument was one of the very early satellite surveying systems. It comprised a 40 lb steel box and 2-metre antenna. It took power from a car battery that needed constant recharging over the 36 hours it took to obtain a position. Days when the Magnavox was running confined two of us to camp, and constant refilling of the generator. The Magnavox 1502b Geoceiver that I used during one season is now in the basement collection of the Scott Polar Research Institute in Cambridge along with some associated photographs and field notebooks. I am not sure that this item has ever made it upstairs into the public museum. This I can understand as it would struggle to capture punters' interest when placed alongside Scott's snow goggles or his final letters home.

By early February the whole Rutford Ice Stream network, some 120 stakes, was surveyed. Professional surveyor, Jonathan was uplifted by Twin Otter and returned to Rothera. Three of us remained to map the thickness of the glacier beneath us, using a system comprising a set of car batteries, a radar transmitter and receiver, an oscilloscope, and a polaroid camera to record the screen of the oscilloscope. It was a cut-down version of a radio-echo-sounding (RES) system that had been designed previously for use in a nice warm aircraft.

Such RES systems are extremely difficult to test in the laboratory. Much of the fine tuning was postponed until we were on site. It was suspected that the thickness of ice we were likely to encounter on Rutford Ice Stream would be much greater than previously measured.

Rick had designed and prefabricated a reflector that would be mounted over an RES antenna directing as much of the radio energy down into the ice as possible. In any context, this reflector would have been a ridiculous structure, but the idea that it might be possible to construct on site and then tow it hundreds of kilometres around the ice sheet seemed preposterous.

Rick Frolich constructing the radar reflector. Mounted over the radar antenna, the frame covered in chicken wire was effective in directing radio-waves down into the ice. In the distance, Mount Mogensen, most northerly major peak in the Ellsworth Mountains. (Credit: David Vaughan/BAS)

The electronics of the RES system that had performed well enough in a warm aircraft proved temperamental rattling over the ice on a cold sledge. At around 2 metres wide, almost 3 metres long, and almost 3 metres high, the reflector resembled a massive triangular tent frame covered in chicken wire. Its spars were pop-rivetted aluminium and the whole thing was mounted on a sledge towed by the skidoo. The sledge rocked and rolled manically, sliding uncontrollably in a crazy fashion. Consequently, we drove at barely more than walking pace for the rest of the season, overheating our skidoos and fraying our nerves, especially, as we were now well past midsummer on Rutford Ice Stream, and the daytime temperatures were routinely close to -30°C. Before each measurement was made the system had to be warmed, and the antenna manually moved up and down in the snow to obtain a good signal. It was hard work, and it took many weeks to make the couple of hundred measurements of ice thickness we needed.

The results of the project we completed in that first season were exciting for us glaciologists. The glacier was around 150 km long, 35 km wide, and in places more than 2500 metres thick[3]. It flowed at up to 350 metres per year. Dynamically, it was contained within a deep trough, and more like an outsized mountain glacier than most of the other ice streams around West Antarctica.

We were due to be uplifted from Rutford Ice Stream in the last week of February. By this time, the twenty-four-hour daylight was waning and for a few hours each day the sun was beginning to disappear behind the mountains. As it dipped out of sight, the temperature plummeted to below -35°C. In the face of such temperatures, our woollen jumpers and cotton windproofs struggled to keep out the cold and having camped since early December we were all dog-tired. Rutford Ice Stream however, had one last treat, a storm of such ferocity that our tent shook and rattled for five days.

Previously, falling snow and poor visibility confined our field party to tents for periods known as a 'lie-up', but this was a proper Antarctic storm. It is not unusual for lie-ups to last several days. Two people sharing a six-foot-by-six-foot pyramid tent, getting through, or even enjoying, a lie-up of more than a couple of days requires a certain mindset. Indeed, it is a test of character under which few could predict their response.

The spatter of blowing snow and gusting wind was like being sand-blasted. The racket was constant. Opening the door-flap to the outside would fill the tent instantly with cold air and drifting snow. Leaving or re-entering required a difficult and time-consuming use of the gap between the inner and outer door-flaps to the tent and made our efforts to brush off the worst of the snow futile. Expeditions out of the tent were limited to the more serious calls of nature, and the retrieval of more food and fuel from the boxes that held our tent down. With the lack of exercise, unending noise, and concern for our safety, sleep was fitful and hard to come by. For several days, we broadcast weather reports to Rothera at three-hourly, and then hourly intervals, as the concern to get us back to the safety of the station grew.

Eventually, the storm subsided, and a couple of red Twin Otter aircraft were dispatched to fetch us. Seeing them appear as tiny dots on the horizon, gave me a huge sense of relief and accomplishment. We had survived the season without major mishap, and had collected some valuable data, which, while they would not change the world, would turn out to be valuable in understanding Rutford Ice Stream and glaciers in

---

[3] - The results of the study we completed in that first season on Rutford Ice Stream were published in 1987 - https://doi.org/10.1029/JB092iB09p08951. It is, however, of some sadness that, even now almost 40 years after the event, this and much of my other scientific work still sits behind a paywall, so somebody is still making money out of it, although it is certainly not the author! Thankfully, much more science is now published in 'open access' publications.

general. Within hours we were uplifted back up the Antarctic Peninsula to Fossil Bluff, where several other field parties were camped out on the ice shelf waiting for onward flights to Rothera.

Over the years I turned out to be rather good at lie-ups, not just as a time to catch up on sleep but as an opportunity to read, think and write. However, I shared lie-ups with some people who struggled with enforced idleness. Some got through it by reading, writing, knitting, drinking tea, smoking, chatting, sleeping or worst of all, humming along to a Walkman. Others simply fretted and stewed.

During one lie-up, my tent-mate, a committed climber, was gravely afeared that his fitness would be compromised by lying on his back for a few days. Each day he stripped down to his thermals and went outside for an hour or two of frenzied digging, returning steaming and ripe. After four days, the storm subsided, and we emerged into a ping-pong ball world completely without contrast. I was unable to see his excavations. It rapidly became clear that each exercise period had resulted in a deep bear trap, and these were dotted randomly round our camp, and were now filled with light blown snow. I tripped over spoil heaps and fell neck deep into several of these before losing my temper and requesting ever so politely that he fill the fucking things in.

Early in 2000, my team (three men and one woman) were caught in a storm on a miserable place called Evans Ice Stream, and apart from one afternoon, we spent 20 unbroken days in our tents. Although I never feared for the integrity of our tents, the unceasing wind was a constant irritation, loud enough to make conversation and sleep difficult. While three of us shared reading materials, my French tent mate did not much enjoy reading in English and so soon ran out of books, and with that any external stimulus. He tried to occupy himself with a task of making a stamp with which to frank our letters home, but after most of a week he had used up all possible materials (candles, pencil erasers, a part of his foam mattress) that were available in our tent. He became silent, listless, and torpid. Years later he said it was one of the most intense experiences of his life – at the time, I had feared for his sanity.

RRS *Bransfield* anchored off Rothera Station, prior to departure for Falkland Islands, March 1986. (Credit: David Vaughan/BAS)

CHAPTER SIX

Homecoming

I had only a short time at Rothera Station before I joined the RRS *Bransfield* to go north to Port Stanley in the Falkland Islands. For a little over a week, I was billeted with a family a few miles out of town at Bluff Cove – a place made famous during the Falklands War for the Argentine air attacks on the Royal Fleet Auxiliary ships, the *Sir Tristram* and the *Sir Galahad*.

Our hosts, Falklands veteran Kevin Kilmartin and his wife Diane, had just had their first child. They were living a tough life as sheep farmers. We camped in their outbuildings. By day, shadowed by Kevin's motley border collies, we roamed across the heath-covered moors that characterise the Falklands visiting notable battle sites. By night we sat by his fireside, talked about Antarctic stuff, and listened to his stories of the war.

Halley's Comet was high in the southern hemisphere night sky, so we woke each night at 02.00, but true to Falkland Islands form, every night was cloudy. I never got to see the comet.

After the Falklands War ended the Conservative government invested heavily in the islands' infrastructure. Top of those investments was a new airport run by the RAF with the dual role of discouraging further aggression by the Argentine military and providing an 'airbridge' from the Falkland Islands via Ascension Island to the UK.

Mount Pleasant airport was fully opened on 1st May 1986, and I was one of the first to use the airbridge. The only problem was that while the airport was ready, the road to it was not fully completed, and we travelled the twenty or so miles on unsealed road on the back of a flatbed truck. Dressed for the tropical stopover in Ascension Island and arrival home, most of us were close to hypothermia by the time we arrived at Mount Pleasant.

Hardpacked, windblown sastrugi. (Credit: David Vaughan/BAS)

CHAPTER SEVEN

Changes in the air; seeing a bigger picture

Perhaps due to long-term fatigue after the six-month trip, I have little recollection of the flight from Mount Pleasant, our arrival at Brize Norton, nor indeed, where I went after I arrived. It is likely that I went back to see my parents in Devon. I do remember that it took several days to readjust to normal life. I had lost various habits we take for granted in the civilised world: locking doors, and carrying keys and money were no longer second nature.

I went back to Antarctica the following autumn not as fresh meat but with seniority, assigned to lead a four-person team revisiting Rutford Ice Stream. I returned many times after that, but never for such eventful trips as that first season, which remained the most significant watershed in my life. In the Antarctic, I had grown up and found a place in the world and felt a little respect coming back to me from a world which itself had changed dramatically while I was south having fun adventures.

In November 1985, Ronald Reagan met with Mikhail Gorbachev. That meeting marked the beginning of the end of the cold war. In January, the Space Shuttle Challenger exploded shortly after it lifted off from Cape Canaveral. All seven crew members were killed, and the most powerful nation in the world was publicly reminded of its fallibility. In April, the nuclear reactor at Chernobyl exploded: a mere 20 months later, a significant concentration of radioactive fallout from the explosion fell at the South Pole.

In the UK, Margaret Thatcher swaggered on as Prime Minister after the Falklands War, but the cracks were beginning to show. The jobless total reached 3.2M, and the newspapers revelled in the political muck around Michael Heseltine's and Leon Brittan's ministerial resignations over the Westland Helicopter Affair.

By November, we had identified the first case of Bovine Spongiform Encephalopathy (BSE) – Mad Cow Disease – a brand new disease entirely of our own devising, born from greed and poor animal husbandry.

Such a list of calamities seemed to galvanise many people to believe that something was going very wrong; humans were damaging the planet in so many ways! Scientists now use the term 'Anthropocene' to describe a new geological period in which human beings were the primary force in shaping change on the planet. Elsewhere too, momentous shifts were beginning in science.

While I was 'South', an international conference of scientists was held in the picturesque city of Villach on the River Drau in Austria. This conference framed the science that would guide the rest of my career. It awoke the world to the issues that would come to dominate science and underpin public discourse for almost half a century.

For the first time experts in everything that was known at the time as 'The Greenhouse Effect' came together. The warming effect that certain gases in our atmosphere have on the climate of our planet, and the idea that increasing the concentration of one of these gases in particular, carbon dioxide (CO_2), would upset and change climate across the planet. The idea was not a new one. Of the many excellent web and book resources on the history of climate science, I was particularly taken by a cutting from a New Zealand provincial newspaper published in 1914 that appeared under the headline, 'COAL CONSUMPTION AFFECTING CLIMATE'. This article read: *"The furnaces of the world are now burning about 2,000,000,000 tons of coal a year. When this is burned, uniting with oxygen, it adds about 7,000,000,000 tons of carbon dioxide to the atmosphere yearly. This tends to make the air a more effective blanket for the earth and to raise its temperature. The effect may be considerable in a few centuries."*

The discovery from a carbon dioxide monitoring station in Mauna Loa, Hawaii that concentrations of CO_2 in the atmosphere had risen by 10% since 1960, was forcing scientists to revisit speculations that human interference could be causing climate warming. Although, distinct in cause and effect, and only tangentially related, the discovery by BAS scientists of the Antarctic Ozone Hole in 1985 had primed the ground for this discussion – if it was possible that the human emissions of one class of polluting gases could cause change at a global scale, perhaps it was possible that a second, similar event could also be occurring.

The report from the Villach conference[4] was measured and conservative, but for anyone who took the time to read it, was chilling. It outlined the main elements of the science and highlighted the possible changes we might expect in climate and sea-level. Qualitatively, it was surprisingly close to the mark. Even more surprisingly, its back-of-the-envelope projections of change were quantitively remarkably close to the changes that we have observed since 1985, and to current projections for the future.

It also contained an important political message. Jill Jäger, an environmental scientist who was one of the delegates in Villach, said the conference turned the idea of future change into a more urgent, and more political issue. Talking to the BBC almost thirty years later, Jäger remembered how she left the event with a feeling that "something big is happening [...] the big adventure here was bringing all the pieces together and get this complete picture and we can see that the changes are coming much faster."

Jim Bruce, Deputy Head of the WMO (World Meteorological Organization), noted that "There was a good deal of corridor discussion, with results being presented by social scientists as well as physical scientists. So, there was an amount of real foment going on, but I am not sure that at the time anyone thought this would be a turning point - but it was!"

The only glaciologist to attend the meeting was Gordon de Q. Robin. Originally Australian, he had been involved in Antarctic science since the 1950s. When I met him in the late-1980s, he was a Senior Research Associate at the Scott Polar Research Institute in Cambridge, and he seemed more stereotypically English than just about anyone I had ever met. He cut a dapper figure in a summer suit and silk tie, and he maintained a certain aloofness that was typical of many English academics. Robin was however, undoubtedly, an excellent scientist who did pioneering work on the interaction of ice and oceans, and in the use of radar to measure ice thickness.

In 1946, pilots flying over Antarctica reported gross and dangerous errors in the radar instruments they used to determine their height above the ground. Even with the plane on the ice, these radars might indicate several hundred metres of 'terrain clearance'. It slowly dawned that the radar waves were penetrating the surface ice and being reflected from something deeper. Twenty years later, Gordon and his colleague, Stan Evans, were successful in developing the Radio-Echo Sounding (RES) technique that turned this weakness into a benefit, deploying a system on an aircraft that could be used to map the thickness of the ice far more rapidly than the other methods available.

---

[4] - https://library.wmo.int/doc_num.php?explnum_id=8512

Gordon was cautious about the likelihood of human-induced climate change, but his Villach predictions for sea-level rise were close to the mark. Our observations of the last 35 years of sea-level have proved them to be remarkably prescient.

The advisory group that was set up after the Villach conference, eventually became the Intergovernmental Panel on Climate Change (IPCC), which to this day provides scientific advice to the United Nations Framework Convention on Climate Change and seeks to limit climate change and its impact on the planet and its inhabitants.

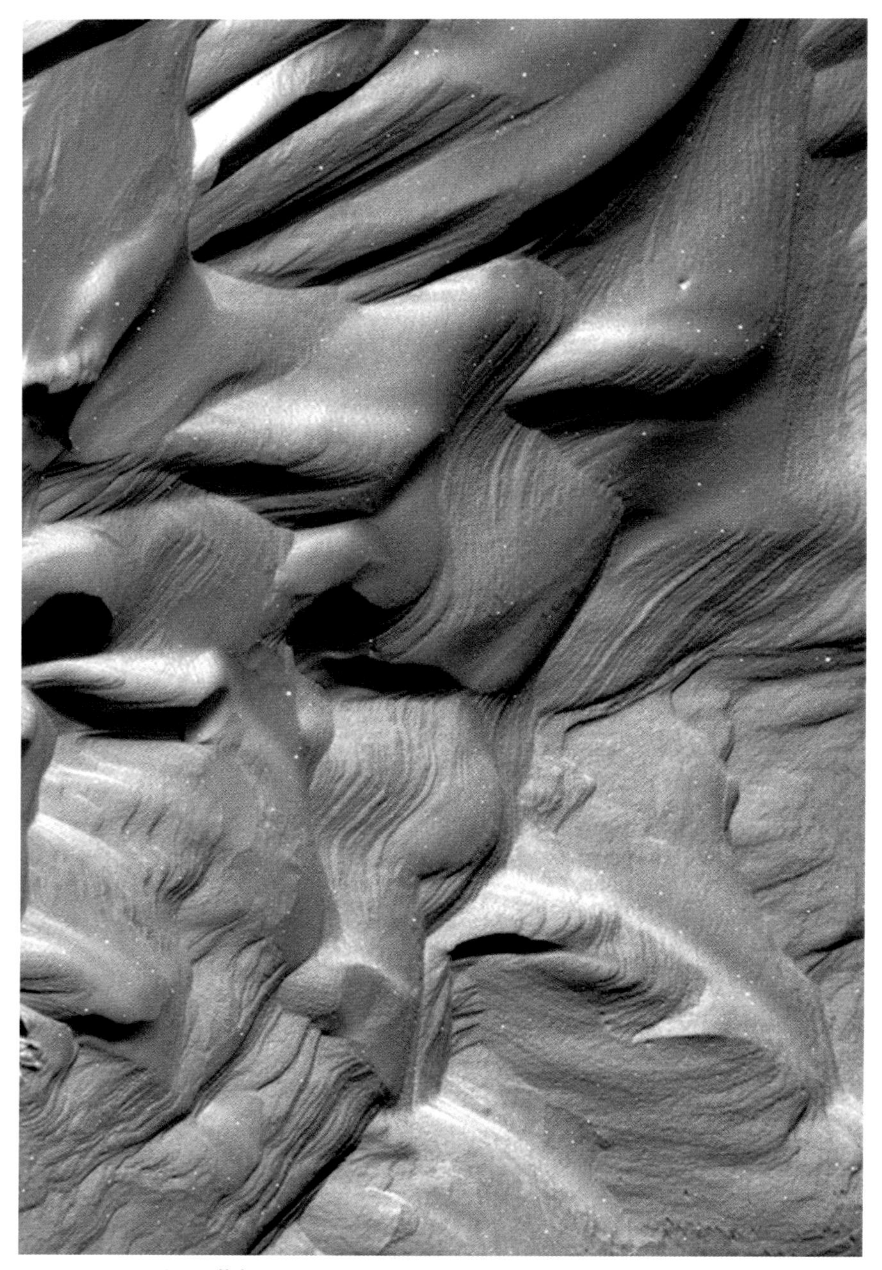

Windblown sastrugi. (Credit: David Vaughan/BAS)

CHAPTER EIGHT

A land you do not see every day

I do not think human beings are born with an inherent appreciation of the beauty of the natural world. Perhaps it is something that is learned or is simply a cultural construct. I remember my parents taking me down to the seawall in Hope Cove, South Devon during one Whitsun holiday, to see the sunset. I can remember wondering what the fuss was all about – I just did not get it. It was only in my twenties that I began to understand and find pleasure in the changing face of landscapes around me.

My entry point was mountain landscapes - mostly, those of Wales, and the English Lake District. What I learned to enjoy was not just their verticality, but rather their grain and structures. These structures bear witness to millions of years of the processes of orogeny and erosion that I learned about in geography and geology classes. For me, it was the naming of the features: arêtes, couloirs, cols, and summits, and understanding how they were formed, that gave me a reason to stare and absorb, and a specific way to relate to them. In the same way that a birder 'collects' sightings of species, or a trainspotter 'collects' the numbers of individual locomotives, I collected good examples of the forms. The superb drumlins visible from the M6 south of Kendal; Crib Goch, the perfect arête on Snowdon; the charming corrie lake Llyn Cwm Llwch that lies beneath Pen y Fan in the Brecon Beacons. And then there was the pleasure to be had in imagining what it would be like to be up there, walking along that ridge, closing that summit, drifting through the clouds.

But as landscapes go, mountains, in all their shambolic, vulgar grandeur, are strident and brash. They are the cocaine of landscapes; and beware, like their chemical counterparts, mountain landscapes can be highly addictive. And what begins as a benign recreation can soon spiral into a problem. For the unfortunate, the longing for mountains accumulates in the blood stream, and in time only stronger and purer doses provide that longed-for hit. A mild, weekend habit, may soon become an obsessive and irresistible compulsion. Some are so compelled by the urge, that little else matters. I have seen friends whose time, money and relationships were consumed by a destructive obsession for the mountains – their lives driven by an unquenchable need.

I was lucky. For reasons I cannot explain, I never got hooked on mountains and remained a 'recreational user' all my life. But beware the power of other landscapes to seduce and beguile. Over the years, I have come to love not just the panoramas of Antarctica but the details, and slowly to appreciate the ice itself, and perhaps this has been one of reasons I was compelled to return to Antarctica so many times. I truly love the ice, its constancy and changeability, and its myriad forms.

Sastrugi

One such form is ubiquitous on the Antarctic plateaus. On a sunny day, even though the air is many degrees below melting, ice crystals will be silently fusing together into a snowpack. With a rising wind, there comes a point when one grain of snow breaks free from its neighbours and is lofted into the air. As it lands it may dislodge others and soon the surface of the snow will be drifting in unbroken mass. Come close to that moving layer and you'll hear the snow singing as grains bounce across the surface and collide in the air. One windy night of drifting snow can change the landscape over hundreds of square miles, creating a million wave-like dunes aligned with the wind. When the wind subsides, a transformed surface remains, a frozen seascape known as *sastrugi*. The word sastrugi derives from a German word which in turn was borrowed from Russian dialect, meaning small ridges or furrows.

Sastrugi may be soft, so you walk through them, or so hard that your boot leaves no mark. They may be a few inches or a few feet high, and can be torture to the travelling field party, causing sledges to slam or slew and rattle themselves to breaking point.

There are some lessons that we seem to have to learn many times over. One such is, not to fight sastrugi but to flex with them. It was the Norwegian explorer and scientist Fridtjof Nansen who, borrowing heavily from Inuit designs, developed a sledge that bears his name and is still in use today. The Nansen sledge is made of ash, beech, twine and rawhide leather, and its secret is that it is designed to bend and flex as it moves. Watching a Nansen sledge carry a load on particularly rough ground is at first alarming; the degree to which it bends, twists and snakes is staggering, but it is this flexibility that gives it the resilience to survive.

Through much of the twentieth century, tractor trains were rarely used for polar travel because their range was limited by the weight of sledge required to tow fuel safely. Many designs of sledge were tested, but all were built to protect a rigid tank containing the fuel. It was only after the light-bulb moment in which that rigid tank was replaced by a flexible rubbery bladder (see Chapter 14) that the capacity/weight equation became workable, and the use of tractor trains became an efficient prospect.

Sastrugi forms dunes resembling a desert-scape in miniature. That balance between order and disorder, like the bark of a tree, or clouds on summer's day, makes them infinitely pleasing to the human eye.

They have an internal structure too. Where surfaces have been scoured away by the wind a rich grain of layers is revealed. Occasionally, one storm will overprint an existing field of sastrugi, leaving a cross-hatched scheme that is almost impossible to decode from the ground, but gain any height and the scheme becomes clear.

Over the years, I have taken hundreds of sastrugi photographs, not for any scientific purpose of classification or measurement, but just to remember the many beautiful textures and form. To those not acquainted with snow, my images are so unfamiliar as to be inexplicable. I have long since abandoned hope of explaining coherently why they move me. It is art, admittedly a low form of art, but I guess you either get it or do not.

Complex pattern of windblown sastrugi, photographed from the air. Two distinct sets of sastrugi are visible, produced by successive storms with different wind directions. (Credit: David Vaughan/BAS)

Ice shelves – special place

Generations of BAS folk have described days where no cloud is visible in the Antarctic sky as 'dingle'. These days are few and memorable. There can be no half measures, no 'mostly dingle', no 'almost dingle', dingle is complete and perfect or not at all. I believe the word derived from some cod-cockney rhyming slang: 'dingle dell, clear as a bell'.

One dingle day in 1992 on the Ronne Ice Shelf, I walked away from the lone tent I was sharing with field guide Crispin. Famously silent and self-absorbed, he was happy to spend a couple of spare hours alone poring over a world atlas and imagining his future adventures. I crunched through last night's soft snowfall. Perfect, featureless, and silent white velvet broken only by the single line of my footprints.

The day before we had driven our little convoy, two skidoos and three sledges out into the middle of the ice shelf to find some buried instruments marked only by a single flag. It had been a clear day. Behind us the Ellsworth Mountains dropped to the horizon, and at 250 km finally passed out of sight. Our overnight camp was hundreds of kilometres from the nearest outcrop of rock, and the same from the icefront - the vertical cliff where the ice shelf calves its icebergs into the Weddell Sea. This was a land of snow and ice, flatter than even the calmest sea.

A couple of kilometres away – our camp was already just a dot in the white – I knelt for a while in the snow. The day was overcast; the sun hidden deep behind featureless cloud, so the sky was the same grey tone in all directions. This ice shelf was faultlessly flat. Through the entire 360-degrees of my view, I could find no single flaw in the fearful symmetry of that empty horizon. No mountain, hill, or hummock. No creature except me to break the silence. Above, the cloudless half-space of sky; below, its mirror image in newly fallen snow. The two meeting in a perfect locus of the horizon.

I have frequently heard ice shelves described, usually by mountain junkies, as 'flat and boring' but I am always hypnotised by their flawless, binary landscape. Their mathematical perfection is incontestable, and for me their simplicity and clarity make them the most perfectly memorable and perhaps the most unexpected landscape to be found on our planet. I am privileged to have visited and spent a little of my allotted time on ice shelves.

PART 2
RETREATING ICE

The rift in the Larsen C Ice Shelf shortly before calving of giant iceberg. (Credit: Pete Bucktrout/BAS)

CHAPTER NINE

The Antarctic Peninsula

Wordie Ice Shelf

After my first Antarctic field season I returned to the continent six times working on worthy, but scientifically niche, projects. The intervening time was spent at BAS HQ in Cambridge. Largely because of my interest and small talent in drawing maps, I became the group's expert in analysing satellite imagery. That sounds more technical than it was. Most of the images were only available as photographic prints, so there was very little we could do with them in quantitative terms. We were limited to recording, classifying, and interpreting them. There were so few satellite images available that I spent hours studying every one of them using a loupe (the kind of magnifying glass typically used by jewellers) tracing paper, sharp pencils, and some long-gone draughtsman's instruments.

Most of the images I used were from the Landsat-4 and Landsat-5 satellites that were launched by NASA, and occasionally from the French SPOT systems. Much of our work was done in collaboration with a German geodesist, Jörn Sievers at the Institute for Applied Geodesy in Frankfurt (IfAG). Jörn was obsessed by geodetic precision to a level that I could not entirely follow, so it was natural that he took the lead in the difficult tasks of rectifying and mosaicking images, leaving my boss Chris Doake and I to interpret the features we could see. Our collaborations produced a series of glaciological maps that were published by IfAG.

These maps were only published after a long delay. There was concern that British Antarctic Survey researchers helping to publish maps with German-language placenames could undermine the UK claim to British Antarctic Territory. It was eventually agreed that a form of words would be printed on the maps to emphasize our spirit of collaboration under the Antarctic Treaty. Permission for me to proceed was granted.

I do not remember if I spotted it, or it was Chris but in 1989 we identified changes to the ice in a series of repeat Landsat-4 and -5 images on the Antarctic Peninsula. These observations turned out to be profound and changed the course of my career. For a moment the eyes of the world focussed on our small corner of Antarctica.

We first noticed the changes in photographic prints, and they appeared so surprising that we took a big step in ordering the magnetic tapes of three images of the same area from 1974, 1979 and 1989. The purchase was a significant dent in our budget, but they arrived from the US in a big box of 12-inch diameter computer tapes, the kind you see whizzing round on the walls of a James Bond villain's high-tech lair. We arranged for the tapes to be sent on to another NERC facility, the Institute for Terrestrial Ecology (ITE) at Monkswood, one of few places in the country where they could be read.

Monkswood was a rather grim piece of 1960s architecture situated a couple of miles from US Airforce base at Alconbury, an hour's drive north of Cambridge. Less high-tech lair more rural business unit. During the next couple of years, Chris and I visited ITE many times to work on the images under the guidance of Roger Parsell, an expert in the use of satellite imagery for farmland classification. Over several weeks, we slowly fixed and warped the images so that they could be accurately overlaid on one another to make a detailed comparison.

Our research paper, published in the journal *Nature* [5] brought scientific and public attention to a series of extraordinary events. Silent changes to the landscapes just a couple of hundred kilometres south of Rothera Station had until now gone unnoticed. Hindsight is a wonderful thing, but by current codes of scientific etiquette, I believe Roger Parsell's technical contribution to this study would have earned him acknowledgement as a co-author of the paper. I am sorry we didn't do that.

In the 1940s, Wordie Ice Shelf was first mapped by the British Graham Land Expedition (BGLE) and named for Sir James Wordie, Honorary Secretary (later President) of the Royal Geographical Society. He had been geologist and Chief of the Scientific Staff of the British expedition, 1914–16, under Ernest Shackleton. Wordie Ice Shelf grew from centuries of ice flowing through the glaciers off the Antarctic Peninsula, captured in a bay, and buttressed by a series of islands.

---

[5] - Doake and Vaughan, Nature, 1991 350, 328–330.

Through the 1960s and early 1970s, it was considered acceptable, if risky, for sledging parties to cross Wordie Ice Shelf enroute to a small, intermittently occupied station called Fossil Bluff. By the late-1970s, the growth of crevasses along this route made it impassable.

Our images revealed that by 1989 the Wordie had roughly halved in area and was criss-crossed by crevasses and rifts that reached through to the sea. Ice rises where the ice shelf had grounded on the seabed had either disappeared completely or remained only as isolated islands – evidence that the ice shelf was thinning as well as retreating. In a few areas where crevasses had not formed, the surface of the ice shelf was dotted with ponds of melt water.

Earlier publications had mapped the ice shelf and even reported some deterioration in its surface condition[6]. Notably, and with a spooky prescience, the reports from the first visitors during 1940's BGLE had already cast doubt on the long-term future of Wordie Ice Shelf. They judged it to be in such poor condition that they believed it was still responding to long-term climate change that began after the Last Glacial Maximum around 20,000 years ago.

Chris and I argued that the changes we observed on the ice shelf did not fit this explanation because losing so much of its area in a little over a decade (1974–1989), was more rapid than anyone could ever have expected. The spectre that more recent changes in climate were to blame became foremost in our minds.

We were aware that the British Faraday research station records showed that air temperatures in the 1960s and 1970s were significantly warmer than those of the late-1940s and the 1950s, but it was a leap at that stage to ascribe this to a warming trend rather than simply a couple of warm decades. The crucial difference being that a trend would likely imply some degree of on-going change, rather than a blip that could be reversed by a few colder years.

For weeks Chris and I discussed the possible underlying cause of the changes in Wordie Ice Shelf. It was clear that the break-up we had observed would not repair itself in a few decades, or perhaps even a century – the flow of ice from the glaciers that fed it was simply too slow to replace the ice that had been lost.

For the *Nature* paper we settled on our conclusion that Wordie Ice Shelf had retreated as a rapid and dramatic response to the much more recent climate warming that had been recorded along the Antarctic

---

[6] - Reynolds, 1988. British Antarctic Survey Bulletin, 80, 57-64.

Peninsula since records began in the late-1940s. It was a bold connection, one that would take more years to verify, but the news was more significant and of wider interest that either of us had expected.

Then as now, science published in *Nature* gets considerable attention within the academic community and the media. While it would be exaggeration to suggest that the disintegration of Wordie Ice Shelf became world news, or had the same scale of impact as, say the Ozone Hole, it was covered in a few newspapers around the world. It was newsworthy because it was the first major climate impact reported in continental Antarctica and was one of the first examples of a large-scale physical system impacted by contemporary climate change, although it was still moot as to whether climate impact could be blamed on human activity.

What our paper had going for it, in terms of news, were the repeated images. These told a visually powerful story so that even non-experts would be struck by the scale of change. The images were reproduced many times, and for a while appeared to be the news editor's go-to visual for the impacts of climate change.

Larsen A and Prince Gustav Channel

While Wordie Ice Shelf was something of a quiet revolution, in the decades that followed changes in parts of Larsen Ice Shelf proved to be more dramatic and more newsworthy, and they began to show that the effects of climate change were being felt even in the most remote regions of the planet. Larsen Ice Shelf sits on the eastern side of the Antarctic Peninsula, running around 700 km north-to-south through a strong climatic gradient from the relatively warm mean annual air temperatures of -6°C in the north to -12°C in the south.

Like so many features in Antarctica, the Larsen Ice Shelf was named by the early explorers who did not have the benefit of the wide perspective afforded us by satellite images. When the ice shelf began to retreat, and its distinct units became disconnected it made sense to give specific names to each unit of the ice shelf. I introduced the subdivisions, Larsen Ice Shelf A to D, in a *Nature* paper by myself and Chris Doake in 1996 and these have been used since.

A far more dramatic change on the Antarctic Peninsula than the break-up of Wordie Ice Shelf took place in 1995. I was in Cambridge when two ice shelves on the northern tip of the Antarctic Peninsula collapsed in a matter of weeks. Together they lost several thousand square kilometres. Both Larsen A and Prince Gustav Channel ice shelves had already been retreating slowly for several years, but satellite images appeared to show a dramatic change in Larsen A.

The low-resolution of these images meant it was not easy to interpret them with confidence. I contacted Rothera Station asking for an aircraft overflight to confirm it. They were reluctant, but as luck would have it there was a planned flight from Rothera over the Antarctic Peninsula to the Argentine Station, Marambio. The route would pass close by Larsen A so the crew was tasked with taking a few photos to keep the 'glacio' in Cambridge happy.

The photographs they sent back showed behaviour never previously recorded. We were astonished; here was something that had never been observed before, and never hypothesised to happen at such speed. And it was acting at such a vast scale that we could see it from space. A 1600-km^2 section of the 200-m thick Larsen-A had in a matter of weeks shattered like a car windscreen, its remains – an armada of football-pitch sized icebergs – now drifting out to sea.

Further north, the overflight confirmed that with the loss of the ice shelf in Prince Gustav Channel, James Ross Island had for the first time in history become circumnavigable. Although at first the channel was still clogged by ice, the Greenpeace ship, *Arctic Sunrise*, claimed this prize in February 1997.

For Chris Doake and me, these latest changes formed a pattern that beyond doubt linked ice shelf changes with regional climate change. Essentially, the pattern of retreating ice shelves fitted our theory that there was a 'climatic limit-of-viability' for ice shelves which had been driven south as climate on the Antarctic Peninsula warmed. Our limit-of-viability had crossed several ice shelves, and these were the ones that had retreated. Once again, we reported our findings in *Nature*[7] and this paper became highly cited as a graphic impact of climate change in action.

What is rarely reported is that two scientists, Helmut Rott from the University of Innsbruck, Austria and Pedro Skvarca from the Argentine Antarctic Institute, were conducting fieldwork on Larsen Ice Shelf in the summer the retreat occurred. Both Helmut and Pedro were reticent to discuss their experiences – both had a long-standing aversion to talking to the media as both had been involved in climbing incidents they thought were badly reported. However, it was probably their experience as alpine climbers that saved their lives as the ice shelf became increasingly treacherous.

Years later the Hollywood film 'The Day After Tomorrow' told a weird story of climate change. Its opening scenes showed a glaciologist on Larsen Ice Shelf as it collapsed. While the CGI graphics of this scene are rather cool, and probably not far from what really happened, it is Helmut and Pedro who were the

[7] Vaughan and Doake, 1996. Nature, 379, 328–331.

inspiration for this scene. I can attest that neither could pass for that square-jawed all-American hero depicted and that neither would have been dumb enough to jump across an opening crevasse.

Antarctic Peninsula hits the headlines

By 1992 BAS was operating a proactive media relations programme led by Linda Capper. We no longer relied on big name journals to publicise our research findings. We were issuing our own press releases and taking media teams to Antarctica to report first-hand what was happening there. The dramatic break-up of Larsen A attracted global media headlines and soon the Antarctic Peninsula came to symbolise our changing climate.

Larsen B

In the austral summer of 2002, the eyes of the world were once again drawn to the Antarctic Peninsula. Within days, almost the entire Larsen B Ice Shelf collapsed producing an armada of small icebergs that drifted off to melt in the Weddell Sea. My friend and colleague, Ted Scambos at the US National Snow and Ice Data Center in Colorado, was ready and his team acquired a spectacular animation of images that showed the change in such graphic detail that it was seen on TV screens around the world.

In barely a month, around 3,250 square kilometres (1,250 square miles) of ice shelf were lost. The area of Larsen B ice shelf loss over the summer of 2002 was a little less than that of Cornwall, and a little larger than Rhode Island. Glaciologists and journalists have made something of a sport of inventing the ridiculous comparative units for the area and volumes of ice-shelves and icebergs. The 'Olympic swimming pool' is well used, as is 'Manhattan', and my personal favourite, 'Wales'.

As the Antarctic Peninsula's ice shelves have retreated, maps have needed revision. Many officially recognised placenames have become misnomers, or entirely redundant. One such was Hektoria Glacier which used to flow directly into Larsen Ice Shelf B; but with the collapse of Larsen B in 2002, the retreating ice left a bay with no name. I am inordinately proud that in 2008 the UK Antarctic Placenames Committee named this new bay, Vaughan Inlet, in my honour.

While the collapses of Larsen A and B ice shelves, were undoubtedly the swiftest, spectacular and newsworthy retreats of ice shelves, similar changes were going on around the Antarctic Peninsula [8].

---

[8] - For a summary, see: Cook, Vaughan and Ferrigno, 2005, Science, 308, 5721

Wordie Ice Shelf has retreated further since the 1980s, the tiny Jones Ice Shelf has all but disappeared, a significant chunk of Wilkins Ice Shelf has gone too.

Working around the world in their offices, and even in their bedrooms, a small army of watchers is now using freely available satellite images to monitor ice shelves with a precision that was impossible to achieve even a few years ago. They have produced captivating animations that help explain the role of fracture in controlling ice shelves. Professor Adrian Luckman from Swansea University has been key in the detailed monitoring of change in ice shelves such as Larsen C for many years.

For me there is sufficient evidence in the pattern and timing of ice-shelf retreat on the Antarctic Peninsula to tie it directly and unambiguously to atmospheric climate change. I do not deny that the ocean beneath the ice shelves might also play a role, possibly in thinning the ice shelves or pre-conditioning them to collapse, but there seems little doubt that there is still strong evidence for the theory of a migrating 'limit of viability' as the primary cause.

However, there is another important lesson we have learned from watching the changes on the Antarctic Peninsula. The loss of ice shelves can have a dramatic impact on the glaciers feeding them. In each case the loss of an ice shelf has been followed by a significant, sometimes dramatic acceleration of the flow in glaciers many kilometres upstream of the ice shelf. This is confirmation of expected theory but is nonetheless important in its implications for other parts of Antarctica where ice shelves are also changing.

The period of strong warming on the Antarctic Peninsula between the 1940s and 1980s, especially in the north, appears to have abated, although temperatures do remain higher than the historical average, and so it is probably unlikely that we will see many more spectacular retreats in this area in coming years, but nor shall we see any rapid regrowth in the ice shelves we have lost – they are essentially gone for good.

PART 3
THE WEST ANTARCTIC ICE SHEET

Rutford Ice Stream. (Credit: David Vaughan/BAS)

CHAPTER TEN

A conundrum

Over and over, I return to the progress, or lack of progress, in the key question that shaped my scientific career. Is the West Antarctic Ice Sheet (WAIS) inherently unstable, or sufficiently close to instability that it could be driven into a phase of irreversible retreat by climate change?

Why, forty years after the question was first posed, is there still no scientific consensus on the central question? Will WAIS collapse in coming decades and centuries fast enough to cause problematic sea-level rise, how much will that sea-level rise be and how quickly will it happen? Why have we so far failed in our mission to reach a justifiable consensus on this? Does that mean we glaciologists are looking for an answer in the wrong place?

I could happily put my own failing in this regard down simply to a lack of innate talent or indeed spending a little too much time enjoying conducting research in Antarctica. However, the failure of one individual cannot have held back an entire research community, especially one of such ingenious, persistent, and often modest scientists as are my many good friends and colleagues. The reason we have not answered this important question lies in a fundamental problem in the question itself. This is not that the question cannot be answered; it is perfectly well-formulated – "well-posed", as my mathematical colleagues would have it. And eventually, with a benefit of hindsight there will undoubtedly be a yes/no answer. The difficulty is in the question's complexity, in the fact that such future ice sheet behaviour will be fundamentally different to all of our current understanding, and indeed, in the assumption that such a future state is predictable with our current knowledge.

This part of the book, whilst recognising that this fundamental question is yet to be finally answered, looks at WAIS and at the early steps of my deep involvement in its research. Taken together, my involvements in WAIS research have helped achieve some of the advances responsible for bringing us closer to the answer. Whilst we do not yet have that final answer, eventually we will and I'm immensely proud of the small part I have played in getting us there.

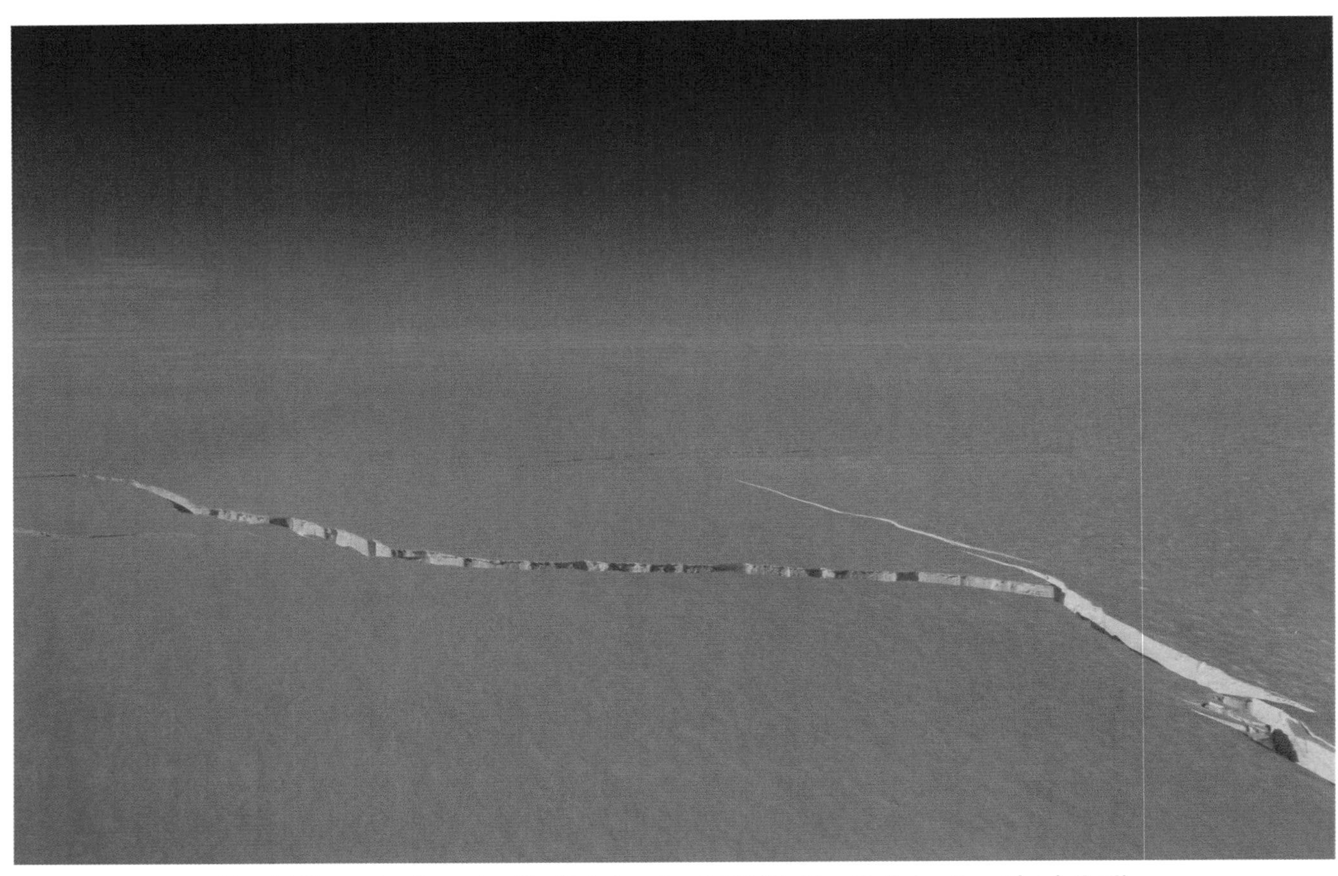

North Rift crack - Brunt Ice Shelf, Antarctica 2020/21. (Credit: Sebastian Gleich/BAS)

CHAPTER ELEVEN

Losing ice

Hypothetically, if the entire Antarctic Ice Sheet melted, global sea levels would rise by about 60 metres. Most of this ice is locked up in the massive East Antarctic Ice Sheet, which because of its size, climate, and the topography of underlying land, is considered 'stable' on timescales shorter than tens of millennia. Elsewhere Antarctic ice is thought to be more vulnerable.

By the mid-1990s, we were accumulating a significant body of evidence to show that ice was being lost from the Antarctic Peninsula. In addition to the retreating ice shelves already discussed, there were several qualitative reports and a few more quantitative studies.[9] I initiated an annual survey close to Rothera Station in 1989 which revealed that the 100-m high snow ramp used as a route to the aircraft skiway had been losing ice at a rate of up to 1 metre per year.

In 2003 I published a paper on how this linked to climate but, it was for others to determine the meteorological factors behind the warming[10]. And given that there is only enough ice on the Antarctic Peninsula to raise global sea levels by at most 10 cm, I was looking towards a larger prize.

The West Antarctic Ice Sheet (WAIS) is the most interesting and scary ice in Antarctica. The volume of ice in WAIS is sufficient to raise global sea levels by up to five metres, and there have long been reasons to believe that this ice could be lost even on timescales of just a few centuries.

---

[9] (Smith et al., 1998. Annals of Glaciology, 27, 113-118). See also: Morris and Mulvaney, 1996. Zeitschrift fur Gletscherkunde und Glazialgeologie, 31. 7-15. Splettstoesser, 1992. Nature 355, 503.

[10] - Vaughan, D.G., et al. 2003. Recent Rapid Regional Climate Warming on the Antarctic Peninsula. Climatic Change 60, 243–274.

In 1978, a relatively unknown glacial geologist called John Mercer published a paper that was to set the course of glaciology for decades to come. Born in England, he was educated in that notoriously tough Scottish private school, Gordonstoun, where early morning runs, and cold showers were compulsory. After serving in the merchant navy as a radio operator for much of WWII Mercer returned to an academic career. His path from the University of Cambridge in 1949, to Canada, Australia, back to Canada, then on to New York and finally to Columbus, Ohio in just 11 years, suggests a certain restlessness and independence of mind.

In Columbus, Mercer became interested in the coming and going of ice in the Southern Hemisphere, including Antarctica and South America on geological timescales. In 1968, he really started putting ideas together. Firstly, he argued that the recent seismic traverses across WAIS, which had shown that much of it rested on rock that was far below sea level, indicated the ice here was potentially vulnerable to change. He reasoned that because the ice rests on rock below sea level, if it began to thin it might effectively peel off around its edges. This reasoning was confirmed by Johannes Weertman in 1974 using a simple ice-sheet model which appeared to show that any ice sheet on a bed below sea level must be unstable and will either grow to fill the continental shelf or collapse completely. I believe this analysis was somewhat flawed because he imagined a 2-dimensional ice sheet resting on a perfectly smooth bed. If added to his model, either of these features – a 3-dimensional ice sheet or more realistic ice sheet bed - could have had the potential to stabilize an ice sheet somewhere in-between Weertman's complete expansion or collapse. However, despite this flaw, the basic idea of WAIS's unique vulnerability to change, remained.

Mercer also found evidence in isotopes that during the previous interglacial period (around 120,000 years ago and a few degrees warmer than the present) global sea level was around 6 m higher than it is today. Mercer postulated that the source of the sea level rise was a massive deglaciation of WAIS, brought about by a warmer climate.

In 1978, Mercer went a step further, suggesting that the magnitude of warming required to cause deglaciation of WAIS was similar to the suggestions made by climatologists about what might occur because of future Greenhouse warming. He wrote:

"A disquieting thought is that if the present highly simplified climatic models are even approximately correct… this deglaciation may be part of the price that must be paid in order to buy enough time for industrial civilisation to make the changeover from fossil fuels to other sources of energy".
J. H. Mercer, 1978 [11].

Throughout the 1980s, Mercer's speculations were explored and formalised by a small avalanche of glaciologists. While his speculations guided and justified the funding of much good work, in honesty, none of it led to any great clarity.

One great proponent of Mercer's theory was glacial geologist Terry Hughes who through the 1980s published several difficult to follow but important works on the emerging discussion around the vulnerability of WAIS. While none of these papers was definitive, or indeed, mathematically sound, each showed a remarkable level of intuition concerning the ice sheet – one might say, a geographer's intuition. And for thirty years Terry had some cachet as one of literally a handful of people, and almost the only glaciologist, who had ever set foot on Pine Island Glacier. He landed there by helicopter from the US Navy ship, *Glacier*, for just a few hours in 1985 with a geologist called Tom Kellogg.

One of Terry's most important contributions was a 'sound bite' that continues to reverberate throughout both media commentary and scientific literature. Terry correctly identified the Pine Island Glacier (PIG), and Thwaites Glacier as the most likely seat of WAIS collapse; he described them as the 'Weak Under-belly of the West Antarctic Ice Sheet'[12].

I met Terry many times over the years. To say he was an unusual character would be an understatement, but he was entirely without malice or side. He marched to the beat of a drum that only he could hear, and he did not much follow the etiquette of modern society. I once observed him at a science meeting in a small hall a few miles outside Washington DC. It was crowded and during the presentations Terry found a place reclining on the floor in the space between the seated audience and the speaker. During one presentation, he got up and shambled to the back of the room where the remnants of lunch were still laid out. He returned with a snack – in each of the four spaces between the massive digits of his right hand he

[11] - J. H. Mercer Nature, 1978. 271, 321–325.

[12] - T. Hughes, 1981. Journal of Glaciology, 27 (97) 518 – 525. This paper is not an easy read, and I have reread it several times to determine the degree to which Terry thought Pine Island and Thwaites were equally worthy of the 'weak underbelly' moniker.

held a chocolate-covered Swiss roll, he had another half-eaten in his left hand – five in all. For Terry, I do not think this was anything strange, and he simply lay down again and began to enjoy his snack.

As a scientist, it would be easy to dismiss Terry simply as a maverick, but I think that would be too easy. He was undoubtedly conviction-led in science as well as in his personal life and honest to the bone. I think there's an important place in research for people like Terry.

In 1998, a paper arrived that finally ended the speculation, giving way to real-world observations. It would shape future research questions surrounding WAIS and its potential contribution to sea-level change.

A team led by Duncan Wingham at the University College London was working on data from the European satellite, ERS-1. This satellite had been operating since 1992 using a radar altimeter which timed the return of a radio-pulse from the satellite from the Earth's surface. After many detailed factors were considered, this highly precise instrument could be used to measure changes in the surface height, and thus thickness, of the ice sheet.

Several other groups had attempted this before, but Wingham's team was particularly adept at correctly applying all the niggly processing steps required to altimeter data (which previous efforts had fudged). They produced a credible map showing change across most of the Antarctic Ice Sheet.

Whilst most of this map showed no strong pattern of change, part of West Antarctica did appear to reveal a coherent signal in the area inland of Thwaites Glacier. Wingham was firm in not overstating his conclusions. He cautioned that it was indeed possible he was seeing a long-term ice-loss on Thwaites Glacier, but that the signal could also conceivably result from a drought in snowfall over the 4-year period of his observations (1992-1996).

Most of us who read the Wingham paper favoured the former explanation and, as more years of data became available, it was confirmed that ice-loss on the Thwaites and Pine Island glaciers could not simply be put down to variations in snowfall. This was the first sign of a negative 'mass balance', and ongoing retreat of WAIS. It was a major step forward and one which caused some embarrassment elsewhere in the science community. The problem was that until this point, those of us involved in fieldwork had been looking in the wrong part of West Antarctica!

Throughout the 1980s and 1990s, two substantial glaciological programmes had developed in West Antarctica. The geographical focus of each was defined more by logistical capabilities of the nations funding them than scientific priority. An Anglo-German initiative (FRISP[13]), working out of Rothera and Filchner Stations, focussed inland from the Weddell Sea, and including Filchner-Ronne Ice Shelf, and the glaciers and ice streams that drained into it, including Rutford Ice Stream. Meanwhile, a US programme working out of McMurdo Station focussed on the Ross Ice Shelf and the ice streams on the Siple Coast. Wingham's result effectively showed that while both these programmes were successful in terms of improving understanding of ice dynamics in WAIS in those areas, when it came to ongoing changes, they were looking in the wrong place. These changes were happening in the third sector of WAIS, the one draining through Pine Island and Thwaites glaciers into the Amundsen Sea. That's where the action was, and we were missing it.

[13] - Originally, FRISP, stood for the Filchner-Ronne Ice Shelf Programme, after the ice shelf in the Weddell Sea, but was later reborn, as the Forum for Research in Ice-Shelf Processes, to mark a widening of interest to include glaciers like Thwaites and Pine Island.

Sheldon Glacier Ice Front taken from Ryder Bay, Adelaide Island Antarctica. (Credit: Pete Bucktrout/BAS)

CHAPTER TWELVE

How a glacier works

A 'glacier' is a term that includes ice sheets, ice shelves, and mountain glaciers. It is defined as: any permanent body of ice in which the ice moves from an area of accumulation to an area of ablation, which removes snow or ice from the surface of a glacier or from a snowfield. Picture a mountain glacier where snow accumulates high up where it is cold, then flows downhill to areas where in summer it melts and runs off into a stream or lake. An ice sheet is similar but here the ice moves from the interior towards the coastal margin where it is lost to melting into the sea and iceberg calving.

The most important concept to understanding change in a glacier is that of the balance between input of snowfall and output of the forms of ablation (melt and icebergs). Glaciologists abbreviate this to 'mass balance', and most glaciers grow until they're close to mass balance, where over a period of time, input and output are the same. However, if one of several external factors change, the glacier or ice sheet will fall into negative (losing ice), or positive (gaining ice) mass balance.

As an analogy, imagine a pile of sand in a builders' yard. New sand arrives on the pile via a conveyor belt that lifts it and pours it down onto the top of the pile. Every day, sand tumbles down from the top of the pile towards the edges where it is taken away by people with spades and wheelbarrows. If the supply of sand via the conveyor belt matches the amount the workers remove, then the volume of sand contained in the pile will remain the same from day-to-day. If the workers take more sand than is falling from the conveyor, the pile will begin to shrink and unless something changes to restore the balance, it will eventually be used up. Similarly, if the conveyor belt speeds up the supply, or the workers take a holiday, the pile will get bigger.

Replace sand with snowfall, and the workers with the processes by which meltwater and icebergs are lost, and you'll get how a natural glacier works, and the concept of 'mass balance' as the factor that controls how much ice is held in the glacier at any point in time.

The concept of mass balance is hugely important because it gives an indication of the state of the ice sheet. Also, since the ultimate destination of any melted ice leaving the glacier is the ocean, it indicates whether the glacier is contributing to sea-level change or not. For a mountain glacier, determining mass balance can be as simple as watching the snout of the glacier retreat or advance, but for an ice sheet covering a huge area, and incorporating complex patterns of flow it is not so easy and involves measuring changes in the ice thickness over wide areas.

Sun pillar and halo generated by falling ice crystals on a cold night
at Pine Island Glacier (Credit: David Vaughan/BAS)

CHAPTER THIRTEEN

Pine Island Glacier

In 1998 with the appointment of a new Director with a background in satellite science a period of real change at BAS began. Professor Chris Rapley was from University College London. Dressed in a sharp suit, and with a very science-focussed agenda, his goal was to bring BAS into the scientific top-tier and address what he called the 'Big Questions'.

At the time, I was feeling somewhat disillusioned with my research career path. In my personal life a breakup with my long-time partner unsettled and shocked me. I started looking for a new role, or even a new job.

With Chris' arrival there was a real sense of opportunity for some of us. We were supported and encouraged to step up into roles that went beyond the confines of Antarctic science.

Moving in on the Amundsen Sea

Monitoring and reporting the ice-shelf retreat and the climate changes around the Antarctic Peninsula had given me a certain scientific profile. It was a nice niche that could have given me an enjoyable and successful career. However, despite the dramatic nature of changes on the Antarctic Peninsula, I felt they were a symptom of climate change, rather than something that would have wider impact. The questions we might address working on the Antarctic Peninsula would always be secondary to those we might answer focussing on the West Antarctic Ice Sheet.

And so, with the impetus of the emerging satellite results and the lack of any competition, I saw an opportunity to refresh the direction of glacier dynamics research at BAS by steering our research over the ridge towards the Amundsen Sea.

Over the ridge to Pine Island Glacier (PIG)

With hindsight this was a no brainer but at the time it was considered, if not revolutionary, bold and a little foolhardy. I set my sights on Pine Island Glacier (PIG). There are no pine trees and no island! It was named by the US, for the USS *Pine Island*, flagship of the eastern task group of US Navy's Operation High Jump which mapped this area in 1946.

There were good reasons why fewer than a handful of glaciologists had ever set foot on this glacier, and why, for two decades, BAS logistics managers had firmly resisted any request to send field parties there.

In the first place, Pine Island Glacier is a very long way from Rothera. With only a few exceptions, Rutford Ice Stream had long been considered the limit for regular field parties, and PIG was about 300 km further away. Using Twin Otters based at Rothera to deploy field parties to PIG was a step too far and was considered too inefficient and risky to make sense.

The weather on Pine Island Glacier was notoriously poor. Atmospheric depressions constantly arrived on the glacier from the Amundsen Sea depositing large amounts of snow. Field parties would have to contend with a lot of digging and any equipment left on the glacier might be buried by several metres of snow before it could be retrieved.

Finally, unlike Rutford Ice Stream, which had been selected as a relatively safe venue for research, much of PIG was covered in lethal and impenetrable crevasse fields.

With these strong reasons not to, we had to work hard to convince our logistics colleagues that work on PIG was both feasible and desirable.

2004/05 with Texas

Our first foray to PIG was in 2004, facilitated by a collaboration with US colleagues who were experts in airborne geophysical survey. Don Blankenship from the University of Texas, Austin had a formidable reputation within the US Antarctic community.

Don's team had developed an excellent suite of instruments that could be deployed by a Twin Otter, to map simultaneously: ice thickness, surface topography, and the gravity and magnetic signals we used to infer the character of the rock beneath the ice. Such data are the starting point for much glaciological research and is often considered an important prerequisite to ice-sheet modelling.

The Texas system was world-leading, and the US Antarctic Program (USAP) managers were rightly proud of its capabilities. They had funded several surveys in remote parts of Antarctica. Given the complexity of the instrument suite, the surveys required a support team of 20-30 scientists and engineers, pilots, aircraft ground crew, 'fuelies', and camp staff.

USAP developed the use of massive fuel bladders, resembling water beds laid out on the ice, to store thousands of gallons of fuel on-site to support aircraft fieldwork. The potential cost and impact of a major leak from these bladders requires a dedicated team to run them. In the Antarctic summer McMurdo station might have around 20 so-called, 'fuelies' working on the station and in the field. See Chapter 14 for more on fuel bladders.

Fuel bladders during the iSTAR fieldwork in January 2014. Teammates taking a rest as the dark bladders soak up some warmth from the sun. (Credit: Jan De Rydt/BAS)

Fuelies deserve special mention because they have one of the toughest jobs in USAP. It is a role that is very often undertaken by women who work almost exclusively outdoors involved in heavy work and much digging of snow. They must take constant care to avoid the instantaneous frostbite that occurs when skin contacts volatile fuel at cold temperatures; and they need to be available anytime, day or night, whenever an aircraft needs to fuel up. They dress in rags because clothes are quickly ruined by fuel and oil. The fuelie motto is "You can smell us coming." Even though their job is grim, they seem to remain cheerful and gregarious. I enjoyed their easy company immensely.

These teams were accommodated on the ice in Jamesway huts. Constructed from a wooden frame covered by an insulated fabric cover with heaters, hard floors, electricity, and rudimentary plumbing, Jamesways became famous during World War II, and then during the Korean War 1950-53. Indeed, much of the camp depicted in the long-running TV comedy, M.A.S.H. was built around Jamesways.

In 2019, I visited one at Lake Bonney, whose maker's mark and date attested that its wooden frame, at least, had been manufactured early during the Vietnam War. They are still widely used by USAP to provide temporary buildings in remote locations. In my humble opinion this is a mistake. Compared to the modern alternatives, they're heavy to transport and require expert carpenters to erect them, slowing down the single-season deployments. However, that said, once up they're marvellous sanctuaries in the very worst weather. Warm, secure, and homey, I do understand why folk like them. BAS never considered Jamesways because they do not fit in the back of a Twin Otter!

While remarkably successful, these were expensive expeditions for USAP to fund, and what pleased USAP managers less than his success in running them was Don Blankenship's delays in publishing or sharing the data his team had collected.

It was not that Don's crew did not want to publish or couldn't see the gems in the data they had collected. Rather, the problem was structural. His team comprised only a couple of full-time salaried research staff and a few engineers. These senior folk spent their time, not on doing the science, but writing funding proposals and supervising an endless stream of grant-funded PhD students; it was these students who were expected to deliver the scientific output. Once a student was allocated a particular dataset, they were given sole access to it. Publication would often wait until they finished their PhD, which in the US might take up to six years. Sometimes, PhD students dropped by the wayside, or did not deliver what was expected and data languished in Austin even longer. The delays in publications increasingly frustrated the USAP community, scientists, and managers. At meetings the tensions were palpable, especially when folk from Texas showed tantalizing slides of the tip-top data they were working on.

Don is not a typical US academic. He is Texan to the bone. Not the 'yee-hah' gun-toting variety, but one of those shrewd, politically astute Texans hailing from Austin – an island of alternative thinking and liberal values, which is often seen as anathema to, and at odds with, the rest of the state. To my English-ear, Don's accent is pure-posh-Southern, marking him out from the standard East-coast US academics that I was used to working with. Don loves argument, intrigue and the cut and thrust of politics. Well-informed on most subjects with more stamina for discussion than almost anyone I know – long after most of us have sloped off to bed, Don would be bringing new points to the table and literally wearing down the opposition. Added to all this, his slight deafness is a real problem in the context of scientific meetings and debates because much of the scientific community is unthinkingly intolerant. In common with a few other deaf scientists I have known, once he had the floor, Don was understandably reluctant to give it up. Overall, the reputation of the Texas group was at something of a low ebb, and the future of their funding potentially under threat.

The airborne geophysical system at BAS was run by Hugh Corr, an enormously capable radar specialist who started his career in Belfast as an apprentice in a company building and servicing commercial lifts. Over 15 years, on a fraction of the budget available to the Texans, Hugh and colleagues designed and built BAS an airborne geophysical system that in technical specification was second only to the Texan one. This was a remarkable feat that entirely justified the Polar Medal that Hugh was awarded in January 2006.

However, in one important aspect our system was superior. It could be operated by a single technician in the aircraft. With the technician also acting as co-pilot, our flight crew was half that required by Texas. By completing a minimum of quality assurance on the data collected in the field, our requirement for scientific support in the field was much lower. This in turn meant less infrastructure and fewer camp support staff. In short, our field party might comprise five to nine people, compared to the 20-30 required by the Texans.

In late 1999, I had meetings with Don Blankenship and the head of US National Science Foundation's (NSF) Head of Polar Programs. The meeting with Don took a couple of days, a lot of coffee and some late nights. The one with NSF took half-an-hour and that included a tour of his house. I was seeking a collaboration in which two temporary field camps would be established on Pine Island and Thwaites glaciers. One, operated by BAS with the aircraft fuel and minimal camp infrastructure provided by USAP. The second on Thwaites Glacier, would be run and occupied by the Texas crew in the normal way. Crucially, my offer to NSF, was that if I was made jointly responsible for the project, I would ensure two things. Firstly, I would do what I could to deliver a logistically-lean field campaign – an approach that was immediately given the slightly menacing nickname, 'BAS-mode'. Secondly, I would commit to agreeing with Don a timeline for publishing the data. I staked my own reputation on a promise that key data would be published within a year of its collection, and that, as soon as they were, the entire science community would have access to them.

Proposals to BAS and NSF were submitted and approved. I visited Austin, and then the Texans came *en masse* to Cambridge to complete the detailed planning. It rapidly became clear that I could not deliver much on my first promise to NSF. The Texans were resolute in their approach to doing fieldwork. Any attempt I made to nudge, push or shame them into 'BAS-mode' was resisted. To be effective they argued that their camp on Thwaites Glacier would need to accommodate around 25 souls, including four pilots who required separate quiet tents with cots to ensure that they would be fresh for duty. It would need several heated Jamesway huts including a mess tent, and one dedicated to science, and an entire village of sleeping tents. Under BAS-mode we planned for a team of nine: a pilot, three camp staff, Hugh Corr and his two engineers, an engineer contracted to operate our borrowed gravimeter, and arguably surplus to requirements, myself as project coordinator and flight-planner.

On 31st October 2004, I stepped out of a C-17 transport aircraft on to the ice at McMurdo Station. I spent several weeks there planning the upcoming survey flights and waiting for our camp on Pine Island Glacier. Considered to be an international airport PIG was designated with a three-letter abbreviation name by the USAP flight controllers. For the rest of the season, we were to be 'PNE' (Pine North-East), or the 'Peonies', to those that got the joke. The US camp ('Thwaites Glacier West', TGW) was abbreviated more conventionally. I joined the rest of my team at PNE on 25th November, and we began work on one of the most significant programmes of airborne survey that BAS ever undertook[14].

Interior of USAP Jamesway hut. (Credit: David Vaughan/BAS)

It should be noted that at the end of the season by flying twice each day, the larger Texas team, did manage to complete more flights than my BAS team did, but I doubt they had as much fun – freezing our bones in our tiny overcrowded and unheated tents at -35°C, we had a great BAS-mode time!

---

[14] https://agupubs.onlinelibrary.wiley.com/doi/epdf/10.1029/2005GL025588 http://www.geo.utexas.edu/courses/387H/Lectures/Holt_06.pdf. For papers that presented data but few conclusions, these are well-cited and influential.

The team at PNE camp on Pine Island Glacier. (Credit: David Vaughan/BAS)

Perhaps one anecdote is worth telling. Fuel was supplied to both PNE and TGW by the Hercules aircraft operated for USAP by the US Air National Guard. They have a long-standing tradition of sending 'combat gnomes' out to far-flung deployments to send pictures and stories home. One day, returning to the PNE mess tent after a delivery of fuel, we found a gnome on our kitchen table. It was a nasty, blue-eyed, leery little thing. We decided that a US military gnome had little place on our British camp. We determined to see him on his way and duly secreted him in the bottom of a box of presents for TGW on the next available flight. A week later the same aircraft, checks done, and engines-running was about take-off from PNE, when a back door opened. The gnome was lowered onto the snow by way of a noose. With the aircraft propellors turning we simply had to watch and accept the donation. Somewhat miffed, we rose to the challenge, and elected to give up a precious box of wine to evict the little sod. Except for one meagre glass, the box's wine bladder was emptied, and enough space made for the gnome inside the box. He was returned with a gift tag and a ribbon. Pleased with our work, we waited but no mention was made in our radio chats with TGW, we assumed that the single glass of wine had been consumed and the box tossed in the trash in disgust as a mean trick. On Christmas Day, a flight arrived from TGW which included a few presents. PNE's primitive cooking facilities did not permit the making of bread and we were collectively bored stiff with BAS-mode biscuits. This massive loaf arriving with the compliments of the chef at TGW was most welcome. It was seeded and golden, and we gathered for lunch in anticipation. Imagine our horror, dear reader, when the bread knife hit hard glaze and the gnome was revealed as it were, cooked-in-a-pie, winking his vicious, victorious wink. We accepted defeat gracefully and he watched over us in the kitchen at PNE until the end. Such is the fun on a field camp in Antarctica!

Air National Guard battlefield gnome emerges from the loaf of bread at PNE camp. Christmas, 2004. (Credit: David Vaughan/BAS)

Having failed to impose BAS-mode logistics on the Texas group, I returned to the UK knowing that to maintain credibility with NSF, it was vital that I deliver on my second commitment – the early publication of our results. After the season, I wrote to Jack Holt and Don and mentioned that I was intending to present results from the BAS side of the expedition at the American Geophysical Union annual meeting in San Francisco in December 2005 and would aim to have a peer-reviewed paper, opening access to the data, submitted by that time. Don was appalled by my reckless timescale and needed to be gently reminded of our commitment to NSF and that BAS would have to deliver on that promise even if his team decided not to match it.

At least part of Texas' unhappiness derived from the fact that presenting our data in their bald form would mean little interest from the high-profile journals, *Nature* or *Science*. We would need to publish in a less prestigious journal. The pressures on scientists to publish in these big journals is enormous, and the credit won in University departments for doing so is significant, but for me the trade-off was worth it and the Texans eventually agreed.

We struggled all summer to process the data but managed to complete the task and presented them to a packed session at AGU (American Geophysical Union annual science conference). A pair of papers were submitted to the journal *Geophysical Research Letters* just before Christmas and both were published early in 2006, releasing data to the circling flock of data-hungry ice-sheet modellers.

In the years that followed, Don's team published many more papers using the data we collected in 2004/5. Several of his PhD students made their scientific debuts on the back of those seasons. Hugh Corr led the publication of a nice paper that showed a nearby volcano had erupted through the ice only 2000 years ago and spread a layer of volcanic ash (tephra) across much of the Pine Island Glacier[15]. We were all happy.

---

[15] - Corr, H., Vaughan, D. 2008. Nature Geosciences, 1, 122–125.

Part of the iSTAR tractor train. The living caboose is towed by a Pisten Bulley tracked vehicle which in turn tows a Lehman sledge loaded with science equipment and supplies, and two fuel bladders, a komatik sledge with explosives, and finally, a Nansen sledge with a ground-penetrating radar. (Credit: David Vaughan/BAS)

CHAPTER FOURTEEN

Game changer

2013-15 – Tractor Train to Pine Island Glacier (PIG)

After our success in 2004/05, BAS logistics relented on supporting projects on PIG, and in the following years we sent several small parties to conduct radar and seismic fieldwork on the glacier itself, led, then overseen, by my dear colleague Andy Smith. While these teams had some significant successes, they all struggled in completing their missions, as they were dogged by difficult working conditions, especially strong winds, cold temperatures, and high snowfall.

During a seismic investigation one of the teams witnessed remarkable behaviour of the glacier. Working 20 kilometres upstream of any crevassed area they used a small hot-water drill to make a series of narrow 20-metre-deep holes into the ice. At half the holes a loud crack was heard. On the surface, a visible crack a few millimetres wide opened in the ice beneath their feet. This crack ran off at high-speed for more than 10 km across the ice. For the team on the ground, this was more than a little scary, but in the years that followed these cracks got wider and longer and became a permanent band of crevasses, each many tens of kilometres long. The small act of drilling a few holes changed the face of the glacier forever – an extraordinary example of the power of tiny disturbances to have huge impacts, the 'Butterfly Effect' if you will.

For completeness, I should also mention a US venture led by Bob Bindschadler. Helicopters were used to establish a small camp on the highly crevassed floating part of PIG. Having drilled through the ice shelf Bob deployed instruments in the sub-ice cavity. A BAS geologist working with the team took sediment cores from the seabed beneath the ice shelf. Analysis of these cores showed that the current retreat of PIG very likely began in 1940s[16].

It was soon clear to Andy and me that the traditional BAS field unit of two people, one pyramid tent, and two skidoos, was inadequate to make best use of the short periods of good weather we saw on PIG.

If we were really going to make inroads in collecting the data we needed then we would need a different way to do it. We campaigned to NERC and BAS to allow us to scale up the operation. It was our dream to bring in use of tractor trains like those that supported the US long-range traverses conducted since the 1957 International Geophysical Year (IGY), but never by BAS.

A typical tractor train consists of powerful vehicles towing lines of heavy sledges stacked with living accommodation, fuel, scientists, and equipment. They have a much greater range and tolerance of poor weather and can support larger teams than skidoos only.

Cautiously, the BAS logistics team listened to our pitch and employed an experienced field guide to visit some of the national programmes who still used tractors. If he had any negative preconceptions, Simon Garrod did not let them show. His season with the US Pine Island Glacier supply team was the game-changer that finally convinced BAS to give tractor trains a go.

Previously, the range of tractor trains was limited by the amount of fuel that could be carried on heavy tanker sledges. It was these tanker sledges that were the problem. The thinking was that they had to be strong and rigid. Consequently, they were heavy. The revolution came in rethinking the need for rigidity.

After much experimentation, the US teams settled on a simple solution of carrying fuel in large, flexible rubberised containers called 'bladders'. Fixing to a half-inch-thick sheet made from low-friction polyethylene plastic provided flexibility when towing large volumes of fuel across the uneven contours of the snow. Crucially, the enormous saving in weight increased the range tremendously.

Over the next year, BAS developed this innovative system for transporting its science teams and equipment over vast areas of Antarctic ice. We purchased two German tracked vehicles of the sort used normally to groom the snow runs at ski resorts. A standard shipping container was converted to provide basic living accommodation, and several bladders were mounted on huge rectangles of polyethylene sheet. The entire

---

[16] - Smith, J. et al. 2017. Nature 541, 77–80.

system was loaded into the hold of the ship and unloaded directly onto the ice within a few hundred kilometres of PIG, along with enough fuel for an entire multi-season project.

The new tractor train was an immediate success, and in November 2013, a party of nine scientists, including me, joined the iSTAR traverse as it came to be called. Another unanticipated benefit was the opportunity for younger scientists with no field experience to join. The increase in the size of field party from two to nine enabled the inclusion of six Antarctic newcomers, mostly from several UK universities. They were supervised closely by us old lags.

The iSTAR traverse was a notable success. A tour of PIG allowed us to gather data that helped us begin to understand the unusual bed conditions beneath PIG, the history of recent changes and the potential for future change. A second traverse followed in 2014/15, and was similarly successful, at which point the vehicles moved on to support other projects in West Antarctica.

PART 4
IPCC

The author, head down, and Tony Payne (Bristol University) during the IPCC WGI Acceptance meeting in Stockholm; September 2014. (Photographer: unknown).

CHAPTER FIFTEEN

Climate change - stepping onto the international stage

In late 1998 there was an open call to join the Intergovernmental Panel on Climate Change (IPCC), which was due to begin preparing its Third Assessment Report. I asked around and none of the climate scientists in BAS were interested, so I sent in a CV. I was selected to join Working Group II, looking at the impacts of projected climate change. As part of the writing team, I was appointed as 'Lead Author', for Chapter 16, Polar Regions. We were a disparate group in the sense of scientific background and nationality but, being unanimously white males over 35, we lacked diversity.

Led by our Coordinating Lead Authors, Oleg Anisimov, a permafrost expert from the Russian State Hydrological Institute in St. Petersburg and Blair Fitzharris (New Zealand), this was exciting in many respects. I learned so much more about the Polar Regions, especially the Arctic. We discussed at length the impact of climate warming on Arctic communities, permafrost melting and causing subsidence of roads and buildings, sea-ice loss accelerating coastal erosion. Oleg reminded us frequently that if you are living in far-north Siberia and your annual heating costs were more than 30% of your entire income, climate warming might be a significant benefit. We discussed the potential for migrating birds to spread disease into northerly ecosystems. I learned from Harvey Marchant, that krill, the iconic species generally accepted to be the foundation of the Southern Ocean ecosystems, were no more significant than the rarely discussed family of zooplankton called salps. Harvey expounded on salps' poorly understood role in the global carbon cycle. Eventually, I asked Harvey what salps looked like. He paused, and in a memorably succinct nugget of cross-disciplinary communication said, "they're basically what's left in your handkerchief after you've blown your nose".

Through 1999 and 2000, we met up around the world – Geneva, Cairns in Australia, and a meeting in Botswana, that I could not attend as my mother was very ill at the time. At each meeting we wrote a new draft for our chapter in preparation for a new round of review. The chapter, one of three, was finally published in the volume, 'Climate Change 2001: Impacts, Adaptation, and Vulnerability'.

That first role turned out to be the start of a long period of service in the IPCC. I was subsequently put in charge of a chapter serving as Coordinating Lead Author (CLA) in Working Group II of the Fourth Assessment Report and Working Group I of the Fifth Assessment. As CLA – a position I shared with one other – I took responsibility to lead the process of distilling the scientific literature into a readable chapter and to take part in the process of distilling the long report volumes into the influential *Summary for Policymakers* (SPM).

Most chapters conform to the rules of grammar but are only 'readable' for the most persistent and tenacious reader. I rarely manage to read more than one section without an overwhelming need to drink coffee.

We CLAs would take that summary through the process of acceptance by the governments involved. For me this final stage was one of the most intense and bizarre affairs in which I have been involved. It is also one which few have ever written about, but it deserves some description. That description follows, but to understand it requires some understanding of the purpose and practice of the IPCC, and particularly the concept of consensus.

IPCC role and consensus, and when it breaks down

The IPCC exists to provide policymakers with regular scientific assessments on climate change, its basis in physical science (WGI); its impacts, implications, and potential future risks (WGII); and potential adaptation and mitigation options (WGIII).

For a scientist who spends every day with the data seeking to affirm the depth of the threat that humans now pose to our planet, the most frustrating feature of the IPCC must also be acknowledged as its greatest strength. This feature, enshrined in the way it was set up, is a lasting testament to the leadership of climate scientists, including Sir John Houghton.

Sir John made sure that the IPCC was owned not by the scientists who write it, but by the governments. In particular, the summary of each scientific report produced by the IPCC is 'approved' line-by-line before publication by those governments. This approval applies only to the SPM. The long-volume scientific chapters serve as the evidence upon which the headline statements are based.

The process of writing the chapters is akin to writing a scientific review, in particular the process of peer-review that is familiar to all who have published papers in high-quality journals.

Coordinating Lead Authors take responsibility for selecting, presenting, and defending the key statements from each chapter in the SPM. These are the most significant and important statements since the last assessment. Before inclusion each must be checked and traced back to the original published scientific papers.

The final IPCC reports are said to represent the 'consensus opinion' of the scientific community. Detractors often cite minority opinions of scientists to dispute this. Consensus within the IPCC process is specific and powerful. There should be a consensus between the Lead Authors (CLAs and LAs) of each chapter on every statement made in that chapter. The reason that writing a chapter requires so much face-to-face time within the writing teams is to ensure that each statement is written, discussed, revised, and agreed by all the authors.

I was guided by this principle throughout the two rounds of IPCC for which I was CLA. If we argued about a particularly divisive point, or one we knew would be dissected further down the line, I would remind the team that a statement would not appear in the final chapter unless we all agree it. The chapters that I was responsible for managed to achieve such agreement, while still making scientifically significant, influential, and newsworthy statements.

One hard-fought and damaging, case was that of Richard Tol; a Coordinating Lead Author on a chapter on the economic impacts of climate change for the Working Group III Fourth Assessment Report. After the final draft of the report had been through its expert and government reviews, a statement was inserted about the aggregated economic impacts: "Climate change may be beneficial for moderate climate change but turn negative for greater warming." This statement, in a leaked draft, was questioned by Dr Bob Ward at Imperial College's Grantham Institute.

The journal that published the first of these papers was persuaded to issue a correction where Tol admitted that 'gremlins intervened in the preparation', and that his analysis was incorrect. Once corrected, the conclusion that low levels of climate change were likely to produce economic benefits was no longer supported. The damage occurred when Prof. Tol made a public statement in Yokohama saying that he had resigned from the writing team of the WGIII Synthesis Report. Tol told the BBC, "The message in the first draft was that through adaptation and clever development these were manageable risks, but it did require we get our act together... This has completely disappeared from the draft now, which is all about the impacts of climate change and the four horsemen of the apocalypse. This is a missed opportunity."

Authors of an IPCC chapter are always aware that their statements will be examined and dissected. A 2010 forensic review of the Fourth Assessment Report (AR4, 2007) commissioned by the Dutch government[17] was well funded and represented a deep dive into the report. The 'assessment of the assessment' was not without criticism, but with an almost audible sigh of relief, allowed Martin Parry (AR4 co-chair of WGII) to say quite rightly to *The Guardian*,

"The conclusions are not undermined by any errors, and we'd like that to be the message the world will take," said Parry. "[They found] a very small number of near-trivial errors in about 500 pages [and] probably 100,000 statements. I would say that's pretty good going."[18]

The review, although generally fair and justified, caused many of us some sleepless nights.

Each statement in the SPM must also be reached by the Governmental delegations. This is achieved during the monumental acceptance meetings which cost a great deal to scientists and governments alike.

The Acceptance meetings

Over the years, I have been to tens of scientific conferences and symposia. I have sat through many talks by scientists of all nationalities and levels of expertise. I have asked more than my fair share of questions, many of which probably indicated my poor level of attention and comprehension in what they had said. Nothing, however, could have prepared me for the acceptance meetings for IPCC AR4 (WGII) in Brussels, 2007, or AR5 (WGI) Stockholm, 2013.

I do not intend any criticism of individuals or groups, or to undermine the process itself. I give my personal view of the extraordinary series of events that take place in the no-man's-land between science and

---

[17] Netherlands Environmental Assessment Agency (PBL). Assessing an IPCC assessment: An analysis of statements on projected regional impacts in the 2007 report. The Hague/Bilthoven, 2010. p.9

[18] https://www.theguardian.com/environment/2010/jul/05/dutch-support-ipcc

governmental policy. These are duels conducted according to prescribed rules of engagement, between teams of passionate and committed representatives doing their best to find lines on which they can agree and can perhaps act. The process is far from perfect, but it works to the extent that there is a defined output that demonstrably has wide support that, at least in theory, cannot be rescinded.

Acceptance meetings allow representatives of the scientific writing teams and the governments that have commissioned them to meet and discuss the line-by-line detail that the scientists cull from the long reports. Those long reports, the bricks, that run to 1,000 pages or more are solely the responsibility of the scientists and do not have to be approved.

Brussels, 2007

I participated in the acceptance meeting for IPCC Fourth Assessment Report for Working Group II in Brussels, 2-5th April 2007. Held in a large auditorium surrounded by glass-fronted booths, these housed a small army of simultaneous translators who worked in shifts to translate the speeches and interventions in English, French, German and Russian.

Scientists sat at the back of the auditorium. In front sat the governmental teams, identified by a little flag on a stick. Some countries were represented by individuals, others by teams of civil servants and their advisors. Seated at a long table on stage facing everyone, were Martin Parry and Osvaldo Canziani, co-chairs of Working Group II, and Jean Palutikov the secretary to Working Group. This team bore the brunt of the work in what would turn out to be one of the most gruelling weeks I was ever involved in.

Suited and booted on Monday morning, we arrived full of anticipation and an excitement to get the job finished. This was the end of a four-year effort by the scientists.

The *Summary for Policymakers* (SPM) would be understood by governments and public, and more importantly would have influence on policy. As Coordinating Lead Authors, Oleg Anisimov from St Petersburg and I represented the scientific team who wrote the chapter on Polar Regions.

I arrived at the meeting at a low ebb. The night before I had taken my wife Jacqui to a snazzy restaurant, The Belga Queen. The former grand old bank was famed for its high-end dining, wonderful décor including a life-size sculpture of a wild-looking horse, and transparent toilet cubicles whose doors only became opaque when locked. I can't imagine the purpose of the final feature except to add a sense of jeopardy to going to the bog and as an obvious 'talking point' for jaded Bruxellois, too bored to be impressed by the first two.

We felt rather special in such a grand place, but as ever my over-ambitious ordering let me down. You might have thought I would have learned a lesson. I started with a medley of eels in duck fat. It did not agree with me, and for the next twenty-four hours I was wracked with diarrhoea and vomiting.

As ever with such meetings, there is a caravan of local dignitaries who need to thank everyone for coming to their fine city, for all our hard work, to wish us all good luck for the week ahead and so-on and so-forth. This took up most of the morning, and work began somewhat anti-climactically after a brief lunch.

Behind the co-chairs, the page of SPM text under review was projected onto a big screen, with the current sentence highlighted. Key scientists were brought onto the stage to answer questions on this sentence, and governments were invited to indicate their acceptance of this sentence by lowering their national flags. If even one flag remained up the sentence was considered to be 'yet-to-be-approved' and we did not move on.

Clearly, the governments were not so bold as to object to statements simply based on distaste, or lack of fit to their national priorities, although many interventions appeared that way. In those instances, long questions to the scientists tested whether sufficient published scientific evidence supported the statement, or whether the statement was correctly expressed in terms of the IPCC's coded jargon around uncertainty. A good many interventions eventually came down to the position of a comma, or the choice of 'some', 'many', 'most', etc. It was dry stuff, and with over one hundred governments represented, it was a slow process. At the end of the first day, when we retired to a reception offered by the city, we were still discussing the first page of the document.

Reconvening at 09.00 am on Tuesday, we continued where we left off. Progress was glacial, especially as many of the headline statements appeared near the beginning of the document. One logjam occurred in discussion about one of the key diagrams that involved three panels. The first showed the projected mean global warming resulting from various scenarios for the future emission of greenhouse gases. These fed into the likely impacts for water, ecosystems, food, coasts, health. It was a clever diagram which would have allowed statements such as, if we emit CO_2 at this rate, these bad things will happen. The problem for the governments, particularly, Russia and Saudi Arabia, was that such a direct connection between oil production and bad consequences was unpalatable. Many hours of push and push-back almost jeopardised business. Many of the potential solutions offered by Martin Parry were rejected. The eventual solution saved face for both sides but truly satisfied no one. The final diagram showed the bad things that were likely to happen due to specific amounts of global warming but did not include mention of the emissions required to reach those levels of warming. These were successfully argued by the recalcitrant governments to be 'climate projections' and so the business of Working Group I, and not Working Group II, and thus outside the scope of this SPM. A complete version of that diagram did appear in the report's Technical Summary, a less-widely read document that was not subject to government acceptance and appeared well after the SPM had been released, reported, and dissected.

On the Wednesday Oleg and I were called to defend the statements from our chapter on the Polar Regions. Some were boiled down so much that in hindsight seem bland. There would be a loss of ice in both Polar Regions and the surrounding oceans with impacts on landscape, marine conditions, and ecosystems, etc. However, after many hours of discussion with Oleg, which had never quite come to the banging of shoes on the table but got close, we had included a couple of statements that raised eyebrows across the green communities.

> *"For human communities in the Arctic, impacts, particularly those resulting from changing snow and ice conditions, are projected to be mixed. Detrimental impacts would include those on infrastructure and traditional indigenous ways of life."*

And,

> *"Beneficial impacts would include reduced heating costs and more navigable northern sea routes."*

In the event, none of our statements provoked much controversy. We spent barely an hour on the stage, then we retired to the back rows to watch the drama unfold.

The document had to be finished by 10.00 am on Friday morning. Martin Parry worked hard to increase the speed of our working and the time available for discussion. He pressed governments to avoid the trivial, and stick to their key concerns, and was short with scientists whose responses to questions lapsed into the realm of lectures. Our Tuesday evening sessions ran until around 10.00 pm, Wednesday ran into the early hours of Thursday. After a short break and a couple of hours sleep and some long hot showers, we reconvened on Thursday morning.

Late on the Thursday night, the entire process teetered on the edge of failure. Utterly exhausted, the translators walked out. Up to this point proceedings were almost entirely conducted in English, but without translation, it appeared that the meeting could not legitimately continue. In particular, the Russian representative who did not speak great English could not follow all that was going on. For a while, it seemed we might all go home with no accepted report. Oleg finally stepped in and offered to move down from the science team to sit with his government representative and provide such translation and, I suspect, moral support, as he could. This placated the Russian representative who eventually agreed to continue.

Our Belgian hosts had planned to give each of us a goody bag at the close-of-play on Thursday, as everyone departed to their hotels with a warm glow to take a well-earned rest and prepare for the Friday press conference. The idea was that we would eventually give our goodies to our families. The centrepiece of these bags was a rather large fine Belgian chocolate easter egg. Having missed the chance, the goody bags were distributed in a coffee break in the early hours of Friday. When we reconvened to complete what would now inevitably be an all-nighter, the resolve of many dissolved, and the eggs broken open. I remember watching one economist sitting in front of me begin deftly removing some gold foil and breaking off just a little of his chocolate egg. Within the hour, he had consumed the entire egg. Around the room, I saw the same was true of many more eggs that suffered an early demise. The resultant mix of exhaustion and sugar-rush had left many heads down on the desks, or simply unable to stay awake. They lolled in the seats while the last few pages of the SPM came under scrutiny. The injection of sugar at this late stage was, perhaps, a masterstroke by the organisers, and certainly reduced the number of interventions. This added to the fact that the final statements consisted of relative anodyne ones around the requirements for the response to climate change to include adaptation and well as mitigation. I do not understand how, along the way, the word 'mitigation' got redefined in the climate context to mean reduction in emissions, which seems to be the opposite of what I expect it to mean, i.e. the reduction of the impacts of climate change, but sometimes you simply have to go with it even if it is perverse.

By early morning, it finally seemed possible that we might get to the end of the document before time ran out.

To my recollection, Martin Parry chaired the entire session expertly from beginning to end. A feat of great stamina and fortitude, for which he earned my respect as a 'old campaigner'. The scientists and government teams were exhausted and dishevelled, except for those from the US government, who looked as fresh and alert as they had at the beginning of the week.

I met many keen minds and good people through the involvement in the IPCC, but none quite won my admiration more than Martin Parry. Robust, vigorous with more than a little military bearing, he sported a significant moustache, a ruddy complexion, and a solid handshake. He was a sailor, a man who clearly spent much of his time outdoors and I saw that he wore the same sailing shoes throughout meetings on several different continents. His academic career spanned many areas, but his natural skill was as a diplomat. He was perfect as co-Chair for Working Group II. He seemed to know everyone, and was warm to every individual, regardless of nationality or background. In that week we all marvelled at his determination and cast-iron constitution, as he patiently steered the discussions to a successful conclusion.

I had voiced concern that planning the press conference on Good Friday morning in a catholic country was unlikely to maximise the attendance. How wrong I was! The press was due to be allowed in at 10.00 am, but as our business ran over, they were held outside. Unhappy to be delayed, once in they made a beeline for the senior figures, Martin Parry, foremost among them.

Most of us scientists, finding we had no more to do, drifted off to a nearby street, where we sat blinking in the Good Friday sun with a jolly Belgian beer. Most of us were stunned by the bizarre piece of political theatre we had just witnessed and were left wondering if this bizarre pantomime was a necessary part of the process.

One Australian veteran explained that the acceptance meetings, and the jeopardy that lies in them, is one of the key features. If we had not reached the end of the process in the allotted time, the SPM, and by implication the rest of our report, would not have been 'accepted' by the governments who commissioned it. We would have walked away without consensus between scientist and governments, and with the shame falling on the nations that had 'obstructed' the process. That potential shame appears, over and again, to be sufficient to rein in the worst excesses of the governments that still see collective action on climate change as a threat to their nations' lifestyles. This jeopardy means that in almost every case the summaries for policymakers from the IPCC assessments are accepted and approved, line-by-line.

Fallout from IPCC 4AR

With Working Group II's report complete, we waited only a few months for Working Group III to publish the Fourth Assessment Report. There followed a period of immense scrutiny, not just of the Summary but also of the full 938-page report. Foremost among these was a review commissioned by the Netherlands government. This review conducted by the Netherlands Environmental Assessment Agency at the request of the Dutch Environment Minister looked at statements in the eight chapters on regional climate change including, my beloved chapter on the Polar Regions. We were required to assist the review, but took no part in it, and merely chewed our nails on the outcome.

When published the review concluded that statements in these regional chapters were "well founded and none were found to contain any significant errors." Martin Parry, responded that the review had found that "the key conclusions of the IPCC AR4 are accurate, correct and supported entirely by the leading science in the field." However, the review brought to light a particular statement in the chapter on Asia concerning the disappearance of Himalayan glaciers. This stated,

> *"Glaciers in the Himalaya are receding faster than in any other part of the world (see Table 10.9) and, if the present rate continues, the likelihood of them disappearing by the year 2035 and perhaps sooner is very high if the Earth keeps warming at the current rate."*[19]

As ever the devil was in the detail. Given the current rapid retreat of many Himalayan glaciers, the statement is undoubtedly true about specific glaciers[20] but taken to apply to all glaciers in the Himalayas, the statement is clearly absurd. The extreme altitude of the high valleys of the Himalayas, and consequent extreme cold climate found in the upper parts of many glaciers, many Himalayan glaciers will likely shrink because of climate change, but it will take many degrees of warming to cause them to disappear completely; certainly, many times more than could be expected by 2035.

Of course, the authors of the chapter included a citation to the source of this statement, a WWF Report, which does indeed include the statement, but passes the buck for its source to another 1999 report, by the Working Group on Himalayan Glaciology (WGHG) of the International Commission for Snow and Ice (ICSI). The actual source of the statement has been hotly pursued, particularly by Fred Pearce for the *New Scientist*[21], but no similar statement seemed to have appeared in any sound peer-reviewed journal. Graham Cogley, a glaciologist from Trent University, Ontario, mocked the timescale of retreat statement saying, "At current melting rates it might take up to 10 times longer". The IPCC's chairman, Rajendra Pachauri simply mocked the entire report as "voodoo science" lacking peer review. The authors of the chapter and statement might have been more generously treated, if they had not been specific in using the term "very

[19] - IPCC AR4 WGII Chapter 10, Section 10.6.2.

[20] For example, between 1985 and 2001, Gangotri Glacier receded at 23 m per year.

[21] https://archive.ph/9UwN6 . The journalist, Fred Pearce has a long history of reporting on climate change issues.

likely" to describe the result – a term that in the IPCC it says means a likelihood of greater than 90 per cent, a rather high degree of confidence in something that will not happen.

If there was a lesson for the IPCC to learn, it was around the danger of citing non-peer-reviewed reports.

Acceptance Redux

In 2013, I found myself at another acceptance meeting with a strong sense of *déjà vu*. In late-September, under leaden skies we gathered in Stockholm for the IPCC Working I, Fifth Assessment Review. I was representing the chapter on the Cryosphere in Working Group I. It was surprising how much this meeting ran along such similar lines and to almost precisely the same schedule as the first acceptance meeting I had attended in Brussels.

My partner this time was a charming, quiet and gentle sea-ice scientist called Joey Comiso. Originally from the Philippines, he obtained his first degree the year I was born. He moved to the United States and worked for NASA in Washington for several decades, leading research in monitoring sea ice largely using satellite data. In his late 70s he was in good health, but the meeting took its toll on him.

By the middle of the week, he was already visibly exhausted, but nothing I could say would persuade him to leave the hall and get some rest. Towards the end of the week, I truly feared for his health. However, he soldiered on and as I recollect made it to the end. While much is said of the intellectual effort put into the IPCC by scientists, little is mentioned regarding the physical and emotional commitment that many scientists, and their families, give to their duties. It should be remembered that all the authors are volunteers, and few are allowed to take time out of their day-jobs to spend on the IPCC duties.

The absolute highlight of the Stockholm meeting occurred around 3 am on the Friday morning, after we had been in continuous session for around 18 hours. In front of the gathered representatives of the 120 governments, and representatives from the author teams, a tiny grey mouse scuttled out onto the stage. He proceeded to centre-stage and paused, looked over his shoulder did a little mousey double-take at the sea of faces across the hall. He froze, stock still for several seconds, as if to say, "blimey, normally I have the place to myself at this time of night". Then I swear he gave a little mousey bow before exciting stage left. For a moment, his appearance broke all tension in the room, and a ripple of delight swept government and scientific teams alike. In the scientific group, someone muttered, "it's Benjy" and most of us got the joke. Fans of the Hitchhiker's Guide to the Galaxy, and there were many present that night, will know that

Benjy was a familiar and appropriate character to appear at that moment. In Douglas Adams' story, Benjy was one of two mice, who were actually "protrusions into our dimension of vast hyperintelligent pan dimensional beings" who had been sent to Earth to oversee a vast psychological experiment that the mice were conducting on human beings. It seemed fitting that Benjy came to see us, on a night we all felt we were part of a grand experiment[22].

How to communicate?

On 12th December 2015, amid a global media hullabaloo, a significant breakthrough in the global response to climate change was reached at the COP21 meeting in Paris. The signing of the so-called Paris Agreement, which came into force 4th November 2016, was undoubtedly a milestone in showing the commitment of the 193 governments that signed and ratified it. Notably, that number does not include two major oil-producing countries, Iran and Libya, and the US withdrew from the agreement under Donald Trump, only to return in the first days of the Biden administration.

The aim of the Paris Agreement was to limit emissions to keep global warming to well below 2°C, preferably to 1.5°C, compared to pre-industrial levels. It was argued that limiting warming to this level would likely avoid many of the worst impacts of climate change. It would reduce the chances of the Earth going beyond 'tipping points' that could cause the worst type of global calamity.

Despite good intentions, seven years on, no plan has been agreed. At each successive COP meeting we are told that although time is running out, we can keep the Paris Agreement alive if we act quickly. However, it is now widely agreed that keeping Paris alive requires a 50% reduction in emissions by 2030. The fact is, that in 2022, both CO_2 and methane levels in the atmosphere are increasing faster than they were in 2015![23] Surely it is now hopeless, and by implication, all the bad things that the Paris Agreement was supposed to avoid are now increasingly likely.

The remaining question is whether there is more traction in accepting that Paris has failed, and that each one of our governments shares the blame for its failure, or to hang on, in increasingly forlorn and potentially dishonest attempts to convince the world that governments haven't failed us quite yet?

---

[22] - Occasionally, Benjy still tweets on climate change as @IPCCMouse. See, https://twitter.com/IPCCMouse

[23] - https://www.noaa.gov/news-release/increase-in-atmospheric-methane-set-another-record-during-2021

I have been asked many times, how do you seek to persuade people who do not believe in climate change. Over the years, my response to this question has changed. Once, I would focus on improving the understanding of the scientific observations and clarity of my description of their significance. Now I am more likely to say, 'screw them, and leave them to it'.

If the substantial weight of evidence both for the reality of climate change, and the human agency in causing it are insufficient to convince some individuals, then they are probably coming from such different starting point, that they will not be convinced by evidence, logic, or any form of rational debate. The 'fault' lies at least as much with them as us.

I do not reserve this approach for any one group. Indeed, there are a variety of groups who for fundamental reasons cannot accept human agency in climate change: religious doctrine, conspiracy theory, or the moral or economic imperative that follows. There are a few simply gunning for a fight, who have a strict personal commitment to the principles of perversity and contrariness. The fact is that these groups have rendered themselves incapable of participation in rational, evidence-based debate, and as such, are beyond any scientist's power of persuasion.

If the starting point in a journey does not topologically allow a route to a particular destination, then it is entirely pointless trying to navigate to it. It is beyond the scientific method to alter a firmly held faith – why try when there is more productive work to be done.

However, I do not mean these individuals and groups are lost, only lost for the moment. It will be no more than a couple of decades before even the most entrenched climate sceptics – a moniker, I use advisedly – will at least be forced to accept the reality that climate change is happening. It will be apparent, painfully apparent, through their own eyes and experience. Sadly, for the determinedly sceptical, there will be no such outcome with regards to the acceptance that climate change is driven by human activities. No matter how much 'evidence' might be amassed there will remain a selection of daemons, deities, and 'weird science', to blame. But their numbers will shrink and in time will become insignificant.

Some colleagues believe fervently that the failure to persuade rests with the approach of scientists, and we should try harder to engage, and make the message more palatable, or at least, more digestible for climate refuseniks[24].

[24] - Here I use 'refusenik' in its derived sense of, 'one who refuses to comply' with an overtone of oppression, rather than the original, 'one who is refused'. https://www.merriam-webster.com/dictionary/refusenik

This is a charitable and generous approach; however, my fear is that it has led to a strict code that appears to guide the reporting of climate change in the media, and the way that scientists wish to be represented. It even seems to be used within the context of the COP negotiations and led to those facile statements year-after-year that time is running out, but if we only act now all may still turn out well.

False optimism supports delay. If we have not yet gone past the deadline, then surely there's still a little time to wait, and argue about the details without agreeing to anything?

Of course, there will be, and should be, a diversity of opinion with scientists taking a view based on their understanding of the evidence, and their own personalities. However, my feeling is that there is an unspoken narrative which most scientists fall back on. For the moment many present themselves as more optimistic than they truly are.

It is clear to me that in coming decades and centuries civilisation will be truly threatened by the inevitable changes in our planet. We have squandered the opportunity we had, to halt the progress of change. In future we must live with consequences that we would not have faced had we acted with resolution even a decade ago. And even now, staring down the barrel, as democratic societies that increasingly reject evidence-based argument, we do not have the tools to deal with issues. Without some resetting of the democratic process our societies are fated to an inescapable descent into a new dark age, indistinguishable from the Dark Ages when the Roman Empire and its civilisation left our shores. Our ancestors succumbed to invasion by new masters who ruled through fear underpinned by myth and unfounded belief.

As a scientist, a rationalist, and a humanist, I have little response to offer. I may rage against the dying of this light, but with the flame of rationalism guttering, what can we do? It is obvious that minimising the amount of climate change will minimise its impacts on us and the planet. But it is fatuous to think that there is any one threshold that we can identify to save us from the key tipping points and state changes that will hit us hardest. While such thresholds may exist for individual tipping points, they will be much more complex than could be described by a simple metric based on global mean warming [25]. With present techniques they will be almost impossible to predict before they have been exceeded. Indeed, it is possible that some (for example: retreat of the West Antarctic ice sheet) have already been exceeded but it will take a while to become evident.

---

[25] - Die-back of the Amazon rain forests is often cited as a major potential tipping point in the climate system but there is clearly no simple identifiable temperature threshold that will initiate its demise. Rather, the threshold may be dependent on a bunch of more subtle changes, such as the degree to which the forested areas are already cut back and weakened by logging etc., and changes in water availability or soil sustainability.

PART 5
THWAITES GLACIER

Aerial view of icebergs in front of Thwaites Glacier 2020. (Credit: David Vaughan/BAS)

CHAPTER SIXTEEN

Big ambition

In the years, that followed the iSTAR Programme, I spent a great deal of effort trying to persuade NERC and NSF to fund a project that might have the scale to begin to answer the big questions around ice-loss from WAIS. I was supported by Professor Sir Duncan Wingham, Executive Chair of NERC and Professor Dame Jane Francis, Director of BAS, and a couple of eminent US scientists. This ambition required a lot of lobbying within the scientific community and the funding agencies.

The programme that emerged was the International Thwaites Glacier Collaboration (ITGC). The name itself an indication of how fragile the collaboration was in the beginning. We spent ages trying to come up with a good name and acronym for the programme but could not agree on the US spelling 'program' or the UK one, 'programme'. After much discussion we agreed on 'Collaboration'[26].

ITGC is and was the largest joint project undertaken by the two nations in Antarctica for more than 70 years – since the conclusion of a mapping project on the Antarctic Peninsula in the late 1940s. The five-year, $50 million programme organised into eight large-scale projects, brings together leading polar scientists who aim to deliver answers to some of the big questions relating to predictions of global sea-level rise.

---

[26] -.See, https://thwaitesglacier.org

Local midnight at LTG on Thwaites Glacier. January 2020. Snow walls built upwind of the mountain tents provide some protection from high winds. (Credit: David Vaughan/BAS)

The programme began in 2018. More than 60 scientists and students conducted investigations into ocean and marine sediments, measured currents flowing toward the deep ice, and examined the stretching, bending, and grinding of the glacier over the landscape below.

On the first of January 2020, I woke to my alarm and light streaming through the thin nylon above my head. The sun was bright enough to warm my little tent and I was starting to get hot in my sleeping bag. Throwing opening the bag, I lazed for some minutes enjoying the peace. For the first time ever, I was on Thwaites Glacier, a kilometre seaward of the grounding line. This was most important location I had ever visited as a glaciologist.

I got out of my sleeping bag, dressed, then emerged from my tent into a flawless blue and white Antarctic morning. Our camp had 15 tents of various styles and functions spread over a few thousand square metres of flat and unbroken ice. I made my way over to the mess tent. I was third in. Catrin our camp manager and Keith Nicholls the senior oceanographer and hot-water driller were in quiet mood, seeing to their chores and relishing the silence. After cursory greetings we settled to enjoy the peace and space of the morning before the others started to stir. The previous night, I had turned in shortly after seeing in the New Year, but apparently a party continued for several more hours. We decided for once, to let everyone sleep.

We were 17 souls on that camp, designated 'Lower Thwaites Glacier' (LTG). Three hours flying from the US National Antarctic Program Camp at WAIS Divide, which in turn was five hours flying from McMurdo station. Even by the standards of my previous trips to Antarctica, this is truly the end-of-the-Earth. Counted in dollars and people's effort, the costs of ferrying people, food, tents, scientific equipment, skidoos, and fuel to this location is incalculable. Our time here is precious. The scientific team will be here for many weeks and must pace themselves; making sure the twenty-four-hour daylight does not lure them to burn themselves out before they have completed their work.

The occupancy of LTG was four more than originally planned, the extra bodies being me, the BBC's chief environment correspondent, Justin Rowlatt, camerawoman Jemma Cox and cameraman Ben Sadd who was filming for the BBC series, Frozen Planet II. Our brief was to record footage of LTG, and share the historic projects being undertaken in this key location.

There was a sense of over-crowding and disturbed equilibrium at the camp. At mealtimes, places to sit were at a premium, and sockets for recharging phones, cameras, iPads and the rest were constantly being re-negotiated.

LTG was something between the traditional US and UK field camps. In contrast to my previous camps the occupants of LTG were a mixed bunch. Mixed in terms of backgrounds and temperaments, and a long way from the old-fashioned 'beardy' field teams I was part of in my early years with BAS. There were some decidedly non-explorer types: an oceanographer who spent a lot of time in his bed, the 'charismatic' leader of an exceptional robotics team, and several folks who were new to remote working and unfamiliar with the intensity of such an enclosed community.

Morning briefing in the Jamesway mess tent at WAIS Divide. (Credit: David Vaughan/BAS)

Newcomers need to work a little to gain acceptance. Entry is smooth and welcoming for those who listen to what they are told about the local setup and rules, get on with their own jobs, and offer their spare time to assist with the camp tasks. Appearing to be idle, spending excessive time drinking tea, or personal over-use of local resources etc do not go down well.

Part of my role on this trip was to smooth the way for the BBC media team through the US Antarctic system and ensure productive relations with the scientific community. I confess that this was a challenge. Wherever we went in those two months, things were stirred up.

The media team was there to tell newsworthy stories to large and demographically diverse audiences[27]. Achieving this objective often created tension among a science and operational community that was tired, strung out about the uncertainties of weather, and under pressure to complete the field season successfully.

[27] See, for example: https://www.bbc.co.uk/news/av/science-environment-52079159

BBC News Climate Editor Justin Rowlatt disembarks from USAP LC130 at WAIS Divide; December 2019. (Credit: David Vaughan/BAS)

Measured solely in terms of the quantity and quality of the media coverage our trip was a great success for ITGC. However, our media team was recalled to McMurdo before the most stressful part of the scientific programme began. While this was a great disappointment to the BBC and me if I am honest, it was the right decision for the successful completion of the programme's scientific goals.

Foremost among those goals, was to drill a 500-m vertical hole through the glacier into the ocean below. A torpedo shaped robot called 'Icefin' was lowered through the hole to explore the underside of the ice and the characteristics of the warm ocean water that melts it. After that, instruments were hung beneath the ice to monitor seasonal and year-on-year changes in the ocean water, to help us understand why Thwaites Glacier is melting so fast. The hole would be drilled by Keith Nicholls and protégé, Paul Anker, using a well-tested but nonetheless temperamental hot-water drill.

Simple in conception, but complex in the detail, Nicholl's drill used kilometres of rubber tubing and a bank of water heaters (or 'burners') to literally piss a hole through the ice with near boiling water. Each burner required a supply of avtur from a 40-gallon drum, a line of which was already standing ready on the ice.

To get the whole thing started required melting ice in a huge floppy rubber tank called the 'flubber', resembling a massive children's swimming pool.

Although the BBC and I departed before the main hole was drilled, we did see the drill in action, making a short pilot hole to test the system and excavate a sub-surface cistern required to recirculate the water. On a fine still morning, the entire team turned out to fill the tank. As the only person on the camp with experience in using an electric chainsaw, I cut out easy-to-manage blocks of snow that could be dragged in a sledge and put into the tank. We moved more than 10 tonnes of ice before lunch. Once the tank was full, the burners were fired up and the water recirculated until it was hot, and then drilling began. With much hissing and plumes of steam, the black rubber drill hose slowly unwound from its reel, passed over a capstan wheel and disappeared vertically down into the snow. The system was working well and after a couple of hours we had reached around 50-m depth. The test was declared a success and the drill removed and burners turned off. Most of us returned to other tasks, and the drillers took a couple of hours off.

ITGC team filling the 10-tonne drill tank. (Credit: David Vaughan/BAS)

Keith Nicholls tending to the burners of his hot-water drill at LTG on Thwaites Glacier, December 2019. (Credit: David Vaughan/BAS)

The big jeopardy of hot-water drilling, which has cost many projects over the years, is that once begun, it is vital that the process continues without pause for however long it might take to complete. With 500-m of ice to drill at LTG this would be something in the region of 36-38 hours, and it would require coordinated manual effort from the entire team. Even a few minutes break in the supply of hot water would likely mean the drill would freeze into the hole and be lost. Once the hole was completed and access to the ocean opened, the team operating the robot would have just twenty-four hours to complete their work, and then finally the long-term instruments would be lowered into place and permanently, irrevocably frozen into position. The final wildcard would, as ever, be the weather. With most of the drill systems ranged across the ice and exposed to the elements, once drilling begins, curtailing it due to bad weather means a huge waste of precious fuel and days spent digging the equipment out. So, they had to pick their moment, and hope for a period of at least four-days without a storm.

The media team was able to film the set-up and testing of the hot water drill and helped with a practice deployment of the robot into the hole. This allowed Justin to get footage of the camp, including some grim footage of him sitting on our makeshift toilet. We were uplifted by Twin Otter aircraft, shortly before drilling was due to begin.

The ITGC media team (left-to-right, the author, Jemma Cox, Justin Rowlatt, Ben Sadd) shortly before departure from WAIS Divide January 2020. Although not anticipated at the time, this would be my final minutes on the Antarctic Ice Sheet. (Credit: David Vaughan/BAS)

The flight out was spectacular. On a day of clear skies, we climbed and circled LTG camp for the last time. The sun glinted off the drill tower, and the figures of the drillers busying themselves with their equipment became smaller and smaller. After 20 minutes of flying over increasingly broken and fractured ice, we began to see a dark shadow of open water on the horizon. Before long, we turned west, parallel to a high, vertical ice cliff, calving icebergs into a dark and oily sea. We flew on, to our right the inland ice, and at its edge a vertical, cliff, perhaps, 70 metres high. Along its length of the ice cliff, wind-blown cornices curled off its top lip, hanging precariously over the sea below. To our left, in the distance, the dark open water of the Amundsen Sea, and in the foreground a tumbling mass of vast icebergs. Each iceberg as individual as a face. Some criss-crossed by crevasses, others having rolled in the water, showing the layering of centuries of accumulation of snow and ice.

Ben shot some of the first footage of the calving of icebergs from Thwaites Glacier – an important record of conditions at a specific point during its retreat – a point which will not be seen again in our lives or those of many generations to come.

It is hard to convey either the beauty or the scale of the scene we saw. Even just the extraordinary intensity of the light reflecting from the snow, standing against the forbidding darkness of the sea below is beyond my power to describe fully. It is, perhaps its impact on me that is the best diagnostic. I was fixed, close and staring through the small plexiglass window unable to drag myself away, only half aware that Justin's increasing bouncing around was beginning to irritate our pilot. The scale of what we could see, is in theory easier. A body falling from the 70-m ice cliffs would spend around 4 seconds in freefall before hitting water below. We flew for almost half an hour along the ice cliff, a similar distance to that from Cambridge to London, past hundreds perhaps thousands of icebergs, each the area of small town. However, these bald figures fail, I think, to give a true sense of the vast landscape that is Thwaites Glacier. It is, perhaps, only travelling through a landscape that can truly give the human mind a true sense of scale. The first astronauts were, I believe, the first to truly grasp the inherent contradictions of our planet's vastness, and its simultaneous pocket-size. In the same way, despite years of study it was that flight over Thwaites Glacier that allowed me to grasp the scale of task we have as glaciologists in understanding and predicting the future of this epic glacier.

Nicholls and Anker were successful in drilling their hole at the second attempt. The first was curtailed due to poor weather, and several days were indeed spent digging the drill out of the snow and setting it up again. The Icefin team was successful in obtaining measurements, samples and most interestingly, video footage of the underside of the ice and the water beneath. Never-before seen images of a strange unexplored world revealed the underside of the ancient ice is so clear that we can see perhaps a metre into it. The base of the ice, sculpted by the ocean, has a delightful, scalloped surface. A few strange creatures it seems find a home here. Most lurk in holes, homes they have created in the base of the ice from where they filter plankton. Below the ice, occasional translucent fish swim slowly in the freezing water. Their internal organs wink as they pump anti-freeze around their frigid bodies. An occasional squid will jet past.

Later in the season, Nicholls' team moved the entire camp to a new site, and drilled a second hole, which too was successful. As a result, we now know much more about the processes that bring warm ocean water into contact with the ice, melting it at more than a metre per year. The long-term instruments gave a year-long record of the influx of warm water.

Their results will help inform and verify the ocean-ice models we use to predict the future of Thwaites Glacier and determine how the melt rate from its underside will change if the temperature of incoming ocean water was to rise, or indeed fall, in future.

Important work

At the start of the season while I was waiting at McMurdo Station to depart for Thwaites Glacier a high-level delegation from the US Government, including some prominent scientific advisors and leaders from the National Science Foundation who were responsible for allocating funding, visited. After a couple of days of being shown lots of cool Antarctic kit and being flown around in helicopters to the best photo-opportunities, this gang was keen to hear about the science. They came to us after a hearty lunch and were in a spikey mood. Having sat through talks focused on the scientific problem and detail of the forthcoming field projects, we faced several very difficult questions, about costs and the allocation of precious resources. Perhaps unfairly, the questions were addressed to the last of our speakers, a relatively junior postgrad who was not just floored as to how to answer but also visibly shaken. It was time to step in. As the lone Brit, perhaps they cut me a little more slack than they did with my colleagues. We needed to refocus the discussion, not on what we were doing or how, or how tough it was going to be, or even what we hoped to achieve, but fundamentally why we were doing it and why they should continue to fund us. I shared some well-rehearsed facts about London, pointing out that a near-identical narrative could be developed for New York, and several other coastal US cities. It was probably one of the most valuable day's work in my scientific career.

Is deep field science expensive?

For a researcher who spent many years working deep-field, living in one tent with one other person, the International Thwaites Glacier Collaboration is a massive undertaking.

In financial terms, the UK and US research funding combined with support from the Antarctic infrastructure is considerable, but for a problem that is undoubtedly a global threat it is modest.

During the latter years of the Second World War President Roosevelt's administration allocated an eye-watering $2 billion for an international team of scientists to assemble, develop and deliver the four atomic bombs produced by Manhattan Project[28]. At the peak of the effort more than 130,000 people were employed. Of course, not all of these were scientists, and it is important to note that the cost of designing and delivering the Manhattan bombs included the industrial effort required to extract the required mass of radioactive nuclides. The numbers are nonetheless equivalent to around $30 billion in 2019.

[28] Atomic Audit: The Costs and Consequences of U.S. Nuclear Weapons Since 1940, by Stephen I. Schwartz

It is true to say that the expenditure on Manhattan was secured not for the benefit of science but for defence. However, I would still argue a strong parallel between the threat posed by the Axis Powers during WWII, and the threat to the entire planet posed by climate change. In both cases, the threat is truly existential, and in both cases, there is a strong requirement for more science to underpin understanding and engineer solutions.

In comparison, ITGC, the largest ever Antarctic programme on sea-level and the impact of climate change was initially estimated to cost around $50 million. While this figure may be something of an underestimate the disparity of the numbers is still shocking. UK Research and Innovation (UKRI) and the US National Science Foundation funded science grants to around $25M, with a similar, perhaps rather higher, cash burden on Antarctic logistical resources, e.g., ships, stations and air operations, provided by US, UK and Korean Antarctic operators.

As a working scientist, I am grateful for the support and for the considerable efforts made by many individuals, especially in the funding agencies in UK and US required to make ITGC happen. I have no doubt that several individuals put their reputations on the line to make it happen, but it is hard not to look around and wonder if we have our priorities right.

Conventional networks of tide-gauges and three decades of satellites data confirm that global sea-level is rising. Our projections indicate that these rates will increase in coming decades because of the combined effects of warming oceans, melting of non-polar glaciers, and changes in terrestrial water storage (in lakes, reservoirs, and groundwater). Together these predictable contributions will be sufficient to threaten many coastal cities, and innumerable coastal communities around the world. However, there remains one unknown which could further increase rates of sea-level rise and increase them dramatically – the contribution from the glaciers draining the West Antarctic Ice Sheet: Thwaites, Pine Island, Smith, Kohler, and a couple of others. These glaciers are already losing ice at rates that are accelerating. They could soon reach a tipping point beyond which the long-term collapse of these glaciers becomes inevitable and irreversible. The future behaviour of these few glaciers represents the major uncertainty in projection of future sea-level and consequently, for the future of coastal populations around the world.

I would agree wholeheartedly that, while the threat of sea-level rise is not the greatest threat to humanity caused by climate change - in terms of the numbers of lives threatened by extreme summer temperatures, the issue of drought in Africa is urgent - but it is hard to see that increased funding for research in this area could ever be seen as an unwise investment.

Funding for ITGC will end around 2025 and it is likely that at that time, our funding agencies will be looking for something 'new' to fund. The engaged, trained, and effective science community developed under and around ITGC will likely disperse and once again compete to seek funding through various dwindling open calls. Continuity, and completion of the task at hand is far from assured.

PART 6
CANCER

Scientist working on a radar in drifting snow. (Credit: David Vaughan/BAS)

CHAPTER SEVENTEEN

Treatment, Covid-19 and retirement

When I came back from Thwaites Glacier Jacqui observed that I had a 'skinny arse', and six months later it was still skinny! Purely on that basis, and a problem with feeling unexpectedly full after meals, a junior locum doctor sent me for tests, and I was diagnosed with cancer. I can thank Jacqui for her concern. Had I waited longer, perhaps…

The months that followed my Stage-3 stomach cancer diagnosis were also the months that Britain was held indoors under restraints imposed by the Covid pandemic of 2020-21. Instead of going to work, we huddled in home offices, spare bedrooms, our kitchens, and outbuildings. We learned to communicate by video conference and through our computers. Strange times indeed, but doubly strange for Jacqui and me. We were constantly on the road to Addenbrookes hospital in Cambridge where I was receiving fortnightly chemotherapy, scans, and endless blood tests.

As the first Director of Science for BAS I worked alongside a Director of Logistics, a Director of Innovation, and a Director of Financial Strategy and Planning. We were appointed around 2014, reporting to BAS's overall director, Professor Dame Jane Francis. I was responsible for 130 scientists and their PhD students. In normal times, this was a role that required a lot of work with people. My time was taken up managing teams, ensuring the motivation of individual scientists, creating the right culture to support the creative process of science, and above all, ensuring that the funding was coming in.

But times were not 'normal', and along the way our little team was diverted by various events and projects that absorbed much time and energy. Among these, were the remodelling of BAS HQ; moving Halley

Research Station when it was threatened by cracks in the ice shelf; helping to design, build and launch the new polar research vessel, the RRS *Sir David Attenborough* (a.k.a. 'Boaty McBoatface'); weathering a couple of financial crises, and managing the impact of Covid on our UK and Antarctic programme.

The role of Director of Science was frustrating and rewarding in equal measure and was not one to be done on a part-time basis. So, in October 2020, I appointed an Interim Director of Science, Dr Anna Jones, to take over the bulk of my duties. It was clear that it would be many months before I had any hope of returning to work with anything like my previous energy, so in the summer of 2021, I retired after 36 years with BAS.

My treatment continued, with the surgery to remove my stomach delayed by the lack of beds available in Addenbrookes. Eventually in April 2021, I went into hospital to have an extended gastrectomy. After a couple of days on a high-dependency post-surgical ward, I found myself in a ward with six other men. I do not know if the pattern was widespread, but three of these chaps, were suffering what I think of as post-lockdown lunacy injuries. After months of forced inactivity, watching endless YouTube videos they had eventually been let out on their bikes and skateboards and seriously damaged themselves. One had broken a horrific number of bones, another had ruptured his liver, and the third his pancreas. None struck me as nutters – just overgrown boys gone a bit mad from being kept indoors too long.

After eight days, I was discharged from Addenbrookes, hoping for the best.

Urn made by the author; summer 2021. (Credit: David Vaughan)

CHAPTER EIGHTEEN

Back in the workshop (December 2021)

In December 2021, the downtime between Christmas and New Year, and 15 months after my diagnosis, I was back in my workshop. I started working with leather in 2010 (see essay 'On Leather') and was making a leather bottle like the one brought up from the mud of the English Channel with the *Mary Rose*, a fighting ship that served in King Henry the Eighth's fleet before sinking off Portsmouth in 1545. While I was searching Google for details I came across the term 'leather bottle stomach'. It seems this is the common name for *Linitus Plastica*, the type of cancer that I have.

The hours spent cutting, shaping, proofing, and working out how the original might have been made, were the very opposite of a time spent in denial. Rather, this has been a time of peace and contemplation. A time to craft a leather bottle and begin my acquaintance with that other 'leather bottle' which has taken root, unseen inside me. An act of true and pure meditation. Time taken hunched over my bench with simple tools: needles, thread, resin, and wax – has been no less significant, no less healing, than they would have been with legs crossed on a mat under a shade next to a tropical beach in an expensive retreat. It has been a time for contemplation and readjustment. I have compared my life with those that ended in harder times, and memories of my teachers renewed.

While I have no idea what will become of this leather bottle I have made, I cannot put the urge to make it simply down to my tendency to gallows' humour. Whatever, the outcome of the coming treatment, my leather bottle might, in theory, last as long as the *Mary Rose* flacket. It is unlikely that it will, but it could, and with grace, perhaps I may still see out my three-score and ten.

Since diagnosis, I had 10 rounds of chemotherapy, and then surgery. I had an extended gastrectomy, which involves the so-called Roux-en-Y procedure, which involves removal of my entire stomach and re-piping of my pancreatic duct into the colon. I then had another two rounds of chemo for luck!

Post-surgery, my recovery was extraordinary. After returning home from hospital, I read, walked the dog, and learned what, and how, to eat without a stomach. The answer, I found is to eat lots of tiny meals and to avoid sweet and starchy things that get absorbed too fast. Failure to do the latter results in 'dumping syndrome', a state not dissimilar to that experienced by diabetics, and sent me to bed for a couple of hours.

Who would have thought that a stomach was one of those 'optional' organs, in the same class as tonsils, wisdom teeth, gall bladder, spleen, etc, that you can get along perfectly well without. Interestingly, appendixes used to be at the top of this list of 'optional organs', but in recent times the purpose of this blind alley off the digestive tract, seems to be finally becoming apparent. The appendix has been identified as a store for gut flora that, during periods of upset are stripped out of the guts along with their contents. Who would have guessed!

I felt increasingly good until mid-October. However, in early October, I went to Swansea University to lecture, and then on to Greenwich, London to help to show off the UK's new polar research ship, the RRS *Sir David Attenborough* before its inaugural journey to Antarctica. In Greenwich, I manned an exhibition stand and talked to visitors at the National Maritime Museum. Three days of talking to folk was good for the soul, but I came home washed out, and suddenly struggling to eat. My weight was dropping off dramatically, and Jacqui joked "you're only half the man I married!" I shot back, that strictly speaking, I am 75% of the man she married, but I took the point, and we requested another scan.

Towards, the end of November, we went back to Addenbrooke's Hospital for the results. There were five of us in the room, which was ominous. Our oncologist, Deirdre, is visibly tearful when she gives us the news. She confirms what Jacqui, and I already knew in our hearts: the cancer was back! I have two new tumours, and these are inoperable. This is not good and pushing Deirdre for a prognosis, we are met with, "with chemotherapy to shrink the tumour, we can keep you going for a while, but we can't cure you. Let's shoot for a year!". I thank her for being so candid, and it was a relief to know why I have been feeling so bad. Being an oncologist is a tough job, and Deirdre and all the team do a great job.

Because in the early stages, stomach cancers show so few symptoms, it is often not diagnosed until it has spread to other organs. For this reason, the outcomes of treatment are not very encouraging. Of the

20,000 people in the UK each year diagnosed with Stages 1-4, almost 50% do not last a year (fewer than 20% diagnosed with Stage4). Five years after diagnosis, fewer than 20% survive (effectively none of those initially diagnosed at Stage4).

If ever you or yours, shows any sign of unexpected weight loss, go to your doctor to get yourself checked out! Do it and do it soon, time is of the essence and early action can change your long-term chances dramatically.

By the end of November, I am back on the chemo. The same routine every two weeks. A blood test, a consultation with one of the 'Onks', then a morning in the day unit with all the other poor sods hooked up to machines that pump the precisely just-safe doses of poison into our veins. I leave fitted with a pump that continues dosing me with drugs for another 48-hours. The dosing is followed by a week of crushing lethargy, accompanied by several other annoying side-effects that you really do not need to hear about. Then I have a few days of respite when I feel better, before the rigmarole begins again.

In truth, I suffer fewer side-effects from chemo than most, and I am thankful for that. However, the time it all takes is dead, and if this is part of our 'year', every week feels like a waste.

On Christmas Eve, I felt suddenly better, the chemo was apparently working and shrinking the tumours, taking pressure off my remaining gut and my kidneys which have been underperforming for a couple of months. Indeed, I felt sufficiently well to continue a tradition that has been in the Welsh side of my family for at least four generations. This is that a special pie is made from the giblets of the Christmas fowl. It is prepared by the 'man-of-the-house' as a late-night supper on Christmas Eve. The meal should really be eaten after returning from the midnight service, but I am no longer Presbyterian and no longer go to chapel. I hang on to the food side of this particular, and some will say peculiar, tradition.

Because it is a great pie, and because I am the last male in my line of Vaughans, I include here the recipe for giblet pie. The making of giblet pie requires more than just following a recipe, it is something of a ritual that should be followed diligently and with due respect. It always begins with a last-minute panic to procure giblets from any or all Christmas fowls. This often involves phoning local butchers, neighbours, etc. To prepare the pie, begin mid-morning on Christmas Eve, but not so early that a glass of something Christmassy and warming produces tutting from onlookers. It is better all round to ensure that other family members are out of the way, and so are less likely to bind on about the blood and guts, and the delicious aroma that rises from your ingredients. Simmer for more than an hour, all the giblets you can lay your hands on - heart, liver, kidneys, crop, neck, in just enough salted water to cover them. Take the liquid for stock and cool the meat. Strip the meat from remaining gristle and cut into bite-sized pieces – take a big gulp before you start, this gets messy! Brown off stewing steak, onions, and garlic in a large casserole dish, then add the giblets and stock, pop this lot in a low oven and leave it there. Around six in the evening, ensure you're sober enough to return to the kitchen. Boil potatoes. Transfer the meat to your pie tin, and cover with a layer of sliced potatoes, these add thickening to the stock and prevent the pastry top from becoming soggy. Top the pie with shortcrust pastry and glaze with egg. Return to the oven until it is cooked. Serve the pie as it is, with strong spirits and, to keep fully with tradition, without vegetables. Ensure a stock of stories espousing the benefits of nose-to-tail eating, the simple pleasures of times past, and enjoy the annual surprise that 'giblet pie' is so much nicer than is sounds! Leave all the washing-up till the morning and retire feeling worthy and fully prepared ready for the excessive over-consumption ahead.

Between Christmas and New Year, I am back in the workshop. This time, for rather obvious reasons, I am making an urn. Again, this is part of getting things in order in my head, facing up to the probability that I have only a short time left.

My urn is a bright green leather cylinder, about 16 inches high and 5 inches across.

After a lifetime in science adhering strictly to SI units, I find myself returning, at least in my workshop, to using inches in preference to millimetres. It is hard to say quite why, but somehow, I find inches and their fractions a more natural unit. Division of an inch into eighths, sixteenths and sixty-fourths is inherently more straightforward than the crude division of millimetres. One quarter of 5 inches, is immediately 1¼", one quarter of the equivalent, 1270 mm is, err, 317.5 mm!

It is hand-stitched in yellow thread and in case of doubt, it has my initials on it. When the time comes, Jacqui will carry it across the fields with friends and as many dogs as can be mustered, to scatter the contents beneath my favourite tree. It is an English oak, which in winter provides dozens of sheep with protection from the rain, and in summer gives them protection from the sun. After that, I suggest Jacqui uses the urn to hide a bottle of the best booze.

It is a little unfortunate that I only checked the recommended volume of urns after I had made it. I found that there's more left of us than I imagined. The recommended volume is apparently 3.5 litres. When measured, my urn is only 1.5 litres. At the rate I am losing weight that may not be a problem and even if it is, I really can't be bothered to make another one. Perhaps, Jacqui could do a little trimming, and limit cremation to my best bits, Afterall, Mr Bastard Tumour is not welcome to join me in the hereafter!

The author working deep-field. (Credit: David Vaughan/BAS)

CHAPTER NINETEEN

December 2022

Notwithstanding Dierdre's prognosis, it is another year on and I am still standing. I have developed a pronounced stoop; I am walking slower and spending most of these winter days by the fire. I have tumours in my liver, bladder, and bowel, and complications are beginning to set in: my poor liver function means I am no longer strong enough for chemotherapy and a prolonged course of steroids is beginning to have noticeable side effects. From here on, treatment will simply be an attempt to reduce the symptoms and preserve my quality of life. But I am still here and once again I am looking forward to Christmas.

It has been a tough year, involving some more chemo and then my participation in the second stage trial of an immunotherapy drug. In the end the trial probably did me more harm than good, as it set off an auto-immune hepatitis that had me in hospital for a week, but no one could have known that when I started out and there's no sense in crying over spilt milk.

A sad part of the progression of my cancer and the inevitable decline in my faculties is the effect it has on Jacqui. It is hard for her to see her husband, her boy, diminish in front of her, and it is curious to see the grieving process begin even before I am gone. Mourning, bereavement, and grief are expected after death of a loved one, but in the case of a partner witnessing a long terminal illness, the process of grief is already underway long before death finally occurs. Psychologists describe this feeling as 'anticipatory grief', and while the experience is quite different for everyone, it is a common response to impending loss.

I have seen something similar when I have been due to go away to Antarctica for a long period of time. In the weeks building up to my departure, Jacqui seemed to get increasingly agitated – almost as if she could not wait for me to leave. Jacqui put the feeling into words, 'It's only once you've left that I can start to look forward to you coming home, so waiting for you to leave is actually the worst bit'. I'm sure that psychologists have a name for this syndrome!

Anticipation of loss can relate not only to the person you once knew, but also the loss of support in the roles and tasks the dying person undertook within the family unit, the loss of financial security, or the loss of the dreams you had for the future.

For Jacqui, it is the last of these that is the cruellest, but she is also coming to realise that there are a lot of things that for almost twenty years, I have done within our partnership, but which I am already unable to do. I have taken quite defined responsibilities for life's tasks. Had the boot been on the other foot, I would have struggled in different things, but just as much as I see Jacqui struggling. I have written down as much as I can in a book to help her out when I am gone.

I feel like I am looking in the face of it now, but for the moment, as Churchill espoused, I keep buggering on.

Epilogue

David died peacefully at home on Thursday 9th February 2023. He always wanted to write a book about Antarctica but was too busy with work. When he started writing this manuscript, he fully expected his surgery and treatment to give him several more years.

Shortly before he died, David asked a few friends and colleagues to review his draft. Like all scientists, he relied on editors and his peers to challenge his words and help him hone his writing. When I said that I would get his book published, he said rather sternly, "Don't promise something that you can't achieve." For once I have proved him wrong.

Those who read early drafts remarked on how much they felt that David was speaking to them. The David in this book is of course rather different to my David, who loved being silly, skipping down the road with his beloved dog and me, and generally having a lot of fun. I miss him.

Jacqui Vaughan

Acknowledgements

I am hugely grateful to those who have helped me bring David's final work to publication.

Our dear friend Chris Aworth masterfully kept our small publishing team on track. BAS friends and colleagues lent their expertise.

I am very fortunate to have Linda Capper, with her expertise and sheer persistence, as Editor-in-Chief. Linda piloted the book at every stage from its original draft through to the final published form with great skill and diplomacy. Her contribution, with the help of several willing assistants, has been immense in making the book the best it can possibly be, restructuring and refining David's text to bring out the power in his story whilst retaining his voice. Glaciologist Andy Smith fact-checked the science and took charge of getting all the colour photos scanned. Journalists and friends Becky Allen and Alister Doyle, together with Chris and Elaine Aworth, Colin McGill, Fran Malarée, Andy Smith and Ian Willox thoroughly read and edited several drafts.

Bonnie-Claire Pickard from British Antarctic Survey's Mapping and Geographic Information Group produced the Antarctic location map to help readers visualise the places David mentions throughout the book. My thanks go also to BAS, SPRI and several of David's colleagues for permission to use photographs from their collections.

Fundraising to enable me to publish was led by Fran Malarée of Churchill College and Colin McGill, whom David met at Churchill when they became lifelong friends.

Additional thanks to the Master, Fellows and students of Churchill College in the University of Cambridge, who have facilitated the publication of this book, and to the following who have supported the publication financially: Colin McGill, Philip Sales, Miranda Wolpert, Catherine Harwood, Toby Perring, Andrew Webster and other generous financial donors.

PART 7
ESSAYS ON ANTARCTICA AND CLIMATE CHANGE WITH DIVERSIONS OF A MORE PERSONAL NATURE

Dad singlehanded in the sailing club's favourite Albacore,
'768' in Tobruk Harbour; July 1963. (Photographer unknown)

Growing up

I was born in 1962, when my parents were living in Tobruk, on the Mediterranean coast of Libya. My dad was working for the UK Meteorological Office, and at that time every RAF station around the world had a small meteorological office attached. My dad had volunteered for this, his second tour of duty in north Africa, and was due to spend three years there. He and my mother had already spent a happy couple of years living an ex-pat life of parties and days on the beach in Benghazi, but with King Idris on the throne in Libya, overseeing the rapid development of the newly discovered oil fields, and seeking to maintain good relations with the UK, Tobruk was an attractive place for them. It offered much scope for fun and a good life, in the compound, on the beach, and on the water.

I have inherited boxes of photographic slides taken during their time in Tobruk. These show a happy good-looking couple in their thirties having a wild time at dances, family parties, on the beach, and out for trips in the desert in their Austin A30. Dad already sported a full beard which he kept until he died, and my tiny mum was tanned and happy.

Mum and Dad in dinner dress in Tobruk. Circa 1963. (Photographer unknown)

After several years of trying, my parents had by this time given up hope of starting a family. Indeed, they had been advised against it. At 34, Mum was already considered too old to be having a first child, and with a history of Crohn's disease and pernicious anaemia she was not in robust health. Nonetheless, she did become pregnant in Tobruk.

As she came to term, the doctor in Tobruk was more than a little concerned and advised that she give birth close to more sophisticated medical facilities. She was flown, with some urgency, to Nicosia, Cyprus. Later, my dad would tell a good story of how, she was taken the fifty-odd miles in an army ambulance with motorcycle escort, to the collection of prefabricated bungalows that constituted the new hospital at RAF Akrotiri. I arrived after a protracted labour, while he was still in Tobruk – not present at the birth, not even on the same continent!

Given my mum's age and health issues, I was and remain, a single child. We returned to Tobruk and completed their tour of duty, then came back to the UK to settle in my mother's hometown of Downham Market, in Norfolk. With the money they had saved, they put down a deposit on a brand-new bungalow from Barker Brothers Builders. It still stands at the corner of the road – modest and unassuming and very much in the 1960s style.

In the years that followed, my parents fed and watered me. I grew and learned to walk and talk roughly on schedule. After a couple of years, we moved to Yateley, a growing commuter village within range of my Dad's new posting at Heathrow Airport.

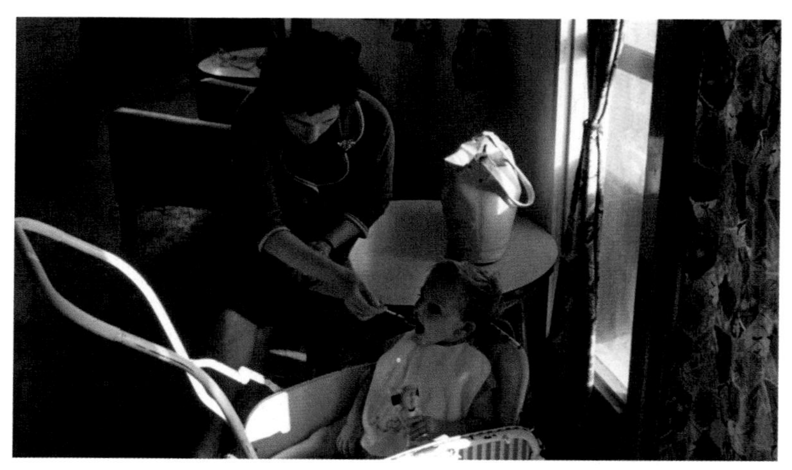

Feeding time at my parents accommodation in Tobruk, Christmas, 1963. (Credit: Garrick Vaughan)

I joined the cubs, went to the local junior school, and grew up. Mum loved me without reservation. Dad maintained some reservations but did a good if stern job of shaping the boy. He taught me how to do stuff in his garage, with bikes, engines, and boats. And though, he would often despair of my lack of common sense – releasing a torrent of reproach, anger, and frustration, from deep inside him that seemed to end only with his exhaustion and utter despair in me – he nonetheless persevered with teaching me to sail. In this, he pursued an old-fashioned syllabus, beginning on dry land with the naming of parts of the boat, and the tying of knots. Then the theory of the sail and aircraft wing, the points of sail and rules of the road. Eventually, when I was six, we rigged the boat and sailed.

I remember, sailing with Dad from Lymington down the river and across the Solent to Yarmouth on the Isle of Wight in our 10-foot Mirror dinghy. I would have been around eight and there was a thick fog. It was an autumn day and we sailed in jumpers, oilskins and plimsols, but the cold crept in. As we left the river, I remember, the land behind us disappeared noiselessly into the fog, and I realized all sight of land had disappeared. We slid quietly into absolute solitude guided only by a compass that swung on a gimbal on the deck. We travelled on into a world containing only our tiny craft and its trailing wake.

I was enchanted by the isolation. The world had disappeared from our view, but more profound, we had disappeared from the view of the world. No one was watching, and no one could see if we were alive or dead. This felt special and exciting.

We arrived in Yarmouth still in fog without a plan. We tied the boat up and walked around the quiet streets while the tide changed, and the blood came back into our hands and toes. On a wet pontoon, we sat and ate shortbread biscuits and drank coffee from a flask with condensed milk, then sailed back the way we came.

When, almost fifty years later, I swallowed my tears to speak at his cremation, it was that day for which I thanked my father. Those few great days shape us more than we know at the time, they set our compasses towards destinations we could never imagine.

I joined the scouts and went to the local comprehensive. I broke an arm roller-skating in our street. I healed and broke it again falling out of a tree in the New Forest. I remember cleaning my bike in the garden of our house during the Easter holidays of 1978 and hearing the first broadcast of the second episode of the 'Hitchhikers Guide to the Galaxy'. I had missed the first episode, and would not hear that for several years, but I was hooked. Being initially confined to radio, the 'Guide' was out of the mainstream, but it had a following even before it appeared on television, and in a less fashionable corner of the playground, a few of us quoted Douglas Adams' masterpiece by heart.

At sixteen, I got a few decent 'O' levels, and we moved again. This time my dad took a posting to RAF Mountbatten in Plymouth, and we found a house overlooking the river in a small village on the coast. I went to another comprehensive on the southern tip of Dartmoor.

Jettisoning the subjects in which I had no skill or interest, I realized that I was quite good at science and maths, and approaching them as sport, I was pretty good at doing exams. Surprising almost everyone, I applied to read Natural Sciences at Churchill College, Cambridge University. My school could not support my taking the Cambridge entrance exam, so I was called for interview. I travelled up on the train and remember being awed by the place and very nervous. In an office with dark wood panelling, I was interviewed by Dr Brian Westwood, a fresh-faced and rather jolly chap who had a strong smell of cologne and was chairman of the Churchill College wine committee from 1977 until 2015. I remember two of the questions he asked me.

First, he asked me to describe the mathematical proof that the square root of two is an irrational number. While I knew what an irrational number is[29], I did not know the proof. I sat and looked dumbly at a piece of paper. I still think this was a tough ask, the proof is a classic, but it was something you either knew or did not, and it wasn't on any syllabus I had studied. In fact, it requires a nasty little piece of mathematical gymnastics called 'proof by contradiction', which did not come up often in school algebra. To this day it is one I can never quite remember quite how it goes[30]. Dr Westbrook was, however, a nice chap, and seeing that I was struggling, and noting from my application that I claimed to be a sailor, framed his second question to give me a chance.

How, he asked, might one reduce weather helm, the tendency of the boat to head up to wind, when sailing a dinghy downwind. This was something I did know. I replied that it was a matter of bringing the centre-of-effort of the boat's sails closer to the centre-of-lateral-resistance of the hull, reducing the length

---

[29] An irrational number is one that cannot be expressed as a fraction of two whole numbers, e.g., p/q, where p and q are integers.

[30] Assuming, the square-root of two can be expressed as a rational number we can write, $\sqrt{2}=p/q$. Squaring both sides of this equation and rearranging gives, $2q^2=p^2$. This means that p2, and thus p, must be an even number. This in turn means q must also be even, and so p and q must have 2 as a common factor. Thus, the initial assumption must be contradicted, and cannot be true. Therefore $\sqrt{2}$ cannot be rational, and it is therefore irrational. This proof is credited to Euclid, who was clearly clever, but he did not get into Cambridge University!

of the lever and the tendency of the boat to come up to wind. I added that this could be achieved either by bringing one sail to the other side of the boat, a so-called 'goosewing' configuration; or leaning the boat to windward, which is both uncomfortable and an unstable way of sailing but achieves the same effect. I guess the second answer clinched it, and I was given an offer to go to Churchill College. On reflection, it wasn't my schooling that got me to Cambridge it was what I learned from Dad, and I am not sure I ever thanked him for that.

Unusually for the time, I was given an offer by Churchill College based on my 'A' level grades, avoiding the Cambridge Entrance Exam, for which many private and grammar schools prepared their students. I was required to get 3 A-grade 'A' levels, and either another A-grade or grade-one in the 'S'-level paper. This was a tough offer, indicating that I wasn't high on their list, but I scraped the minimum requirement, and duly arrived in Cambridge in October 1981. On the journey from Plymouth to Cambridge, with Mum feigning sleep in the back of the car, Dad gave me the obligatory instructions for a lad entering a likely den of vice. I was instructed not to take drugs, and not to get anyone pregnant, and if I did either, not to bother coming home!

As Cambridge undergraduates, we lived a gilded although far from extravagant life – not quite the one depicted in Brideshead Revisited, which was the hit TV serial of 1981, and which coloured everyone's view of life as an Oxbridge undergraduate. There was a punishing schedule of lectures, lab work, and supervisions, but there was also time for sport, parties and learning something about sex.

For the first time I discovered enjoyment in reading. I trod a predictable path through, Vonnegut, Schaffer, and more pretentiously, Voltaire, de Beauvoir and Anaïs Nin. We partied, we posed, played, we loved and got pissed. In the summers, I went back to Devon and taught sailing, and spent a grim month with an interrail ticket touring sights and galleries through France, Spain, Italy and Greece.

For the first time I saw bands. In my first term, I descended a single staircase, in skinny black jeans and eyeliner, into a notorious firetrap called the Sound Cellar, to see Wreckless Eric – the band were very late on stage because the "transit van broke down"[31].

Undergraduate Natural Science courses in Cambridge were run under a 'Tripos' model, where in theory, a student could change discipline from one academic year to the next. In practice, it was not so easy, as a much of the syllabus built on knowledge gained year-by-year. At the outset, I was intent on following the courses that led to a final year studying Physics and Theoretical Physics, but in the first year I really did not shine.

In summer 1982, while I waited for the results of my first-year exams to come through I was helping deliver a yacht from the UK to Gibraltar, with a chap called Jimmy Allen, and my best friend from school, Andrew Jennings. It was a truly horrid passage across the Bay of Biscay during which only one narrow berth in the yacht remained dry. Andrew and I shared this top-to-tail for six days. Arriving at Vigo on the Northern coast of Spain, I went ashore with a small pile of pesetas to hand. I called home from a call box next to the yacht club. Dad had opened the letter containing my results but was confused by the grading scheme. He simply kept saying, 'you've either done very well or very badly'. It turned out it was the latter; I had achieved thirds, a scrape above 'fail', in every subject.

When I finally, returned to the UK, I girded my myself and returned the call from my Director of Studies. Archie Howie was an eminent but dour physicist with a thin face and a strong Scottish accent. Professor Howie firmly tried to dissuade me from continuing with physics, 'have you ever thought of metallurgy, laddie?' Indeed, I had never thought of metallurgy, and wasn't going to begin thinking about it then, but Archie had a point, I was not good enough at maths to ever really cope, let alone excel, at theoretical physics.

---

[31] I have the clearest memory of crushing regret that I missed the Stranglers playing at the Corn Exchange in that first year. This would have been one of the final concerts before the Corn Exchange closed for a refurbishment that took many years to complete. I clearly remember passing by the Corn Exchange as the audience spilled out, steaming onto the wet streets, envious that I had not been inside. It turns out, however, that this is an entirely false memory. By the power of the web, I find that the Stranglers played the Corn Exchange only in March 1981, a full six months before I went to Cambridge. Strangely, this false memory has become embedded, and I struggle not to remember it – it does make me wonder what else I've made up.

While it is a sweeping generalization that surely fails in many individual cases, there is a fundamental difference between a good engineer and a good physicist, which was inherent in how the courses were taught in the 1980s.

An engineer brings to any problem strong intuitions, and then uses a mathematical description to quantify that intuition. That approach is one that may serve a physicist well enough for a while. They may cope with basics physics (Newtonian mechanics, fluid mechanics, etc), even with theories that stretch and require some re-wiring of their intuition (e.g., special relativity, and even some aspects of quantum mechanics). However, as theoretical physics reaches its higher planes, the physicist must cast aside all prior intuition, and truly understand the physics as a necessary consequence of the underlying maths. That tricksy proof by contradiction becomes a trivial party piece compared to the maths underpinning second and third-year physics.

In truth, I was lost completely at the point where our syllabus lost its reference in the real world that I could touch and see and disappeared down a rabbit hole of maths. Specifically, late in the second year with the introduction of Hamiltonian mechanics[32] and ladder operators[33].

After that, I struggled with lectures and course work, and only really did well in laboratory work. Through stubbornness, and a lack of options, I kept going with the final year of Physics and Theoretical Physics, and eventually got a lower second-class degree. I was disappointed, although not surprised, but this mediocre grade closed the door to most of the research opportunities I had imagined myself following. In desperation, I asked Archie Howie to investigate whether I had been harshly marked or if there was reason to appeal my grade. He came back a couple of days later and told me very firmly, I was lucky to have got a lower second, and I shouldn't rock the boat.

---

[32] - I rest my case with the Wikipedia entry for Hamiltonian Mechanics… Hamiltonian mechanics are a reformulation of Lagrangian mechanics where the generalized velocities are replaced with generalized momenta. Hamiltonian mechanics have a close relationship with geometry (notably, symplectic geometry and Poisson structures) and serves as a link between classical and quantum mechanics … [and breathe!]

[33] - In the same vein… In linear algebra and its application to quantum mechanics, a raising or lowering operator (collectively 'ladder operators') is an operator that increases or decreases the eigenvalue of another operator. In quantum mechanics, the raising operator is sometimes called the creation operator, and the lowering operator the annihilation operator. Well-known applications of ladder operators in quantum mechanics are in the formalisms of the quantum harmonic oscillator and angular momentum… [WTF].

Replica of the Marie Rose flacket made by the author in September 2020. (Credit: David Vaughan)

On leather – some history

Two decades ago, I moved from Cambridge to rural Northamptonshire to live with my new wife, Jacqui. Northamptonshire is an under-appreciated county, the very heart of England. It has a supporting role in English history, and its green and pleasant, but unspectacular countryside means it is not on the tourist trail and rarely featured in the where to go guides. But there is one area of endeavour in which my adopted county leads all of England, arguably the world, and this is tanning leather and the manufacture of fine leather goods.

From the 1700s through to the twentieth century, the banks for the river Nene were home to factories that tanned animal hides and sent them into town where craftsmen cut and shaped, and stitched and burnished leather. It was here in Northampton that made some of the finest leather goods available.

Through the nineteenth and twentieth centuries, handmade shoes in particular, the highest achievement in the leatherworker's art, were produced by companies that were household names, at least in the most affluent homes. Some of these companies, such as John Lobb founded in 1866, and Crockett & Jones founded in Northampton in 1879, are still selling shoes today. The firm Joseph Cheaney and Sons have made shoes on the same site in nearby Kettering since 1880, and it is still run by the same family.

These companies still produce some of the best shoes in the world, and despite the struggle to survive in the modern age of fads and cheaper imports, they get by on their customers' desire to own and wear something with heritage. Something that will last. Something that's been shaped and given life by men and women whose craft was learned through years of practice. Admittedly, many of these companies are now owned by multinationals, but some manufacturing remains in and around Northampton, and the craftsmen live among us – a source of pride across the county.

I had an uncle, Morris, brought up in the fens of Norfolk, who moved to Northampton in the 1960s, to work in the shoe trade and bring up a family. He sold Church's shoes to the trade, travelling all over the country to sell to shops and stores, but driving home almost every night to sleep in his own bed. He was proud to work for Church's.

I own several pairs of Loakes shoes and am proud each time I walk out in them, an inch taller, but rooted better in the soil of my adopted home.

Arriving in Northamptonshire, the history of the leather interested me. In 2010 I accidently came across a shop that still sold leather craft supplies to the trade and public. I wish it was a dusty emporium down a quaint street, but it was a warehouse on a small industrial estate. It closed within a year but was full of tools whose purpose I did not know, dyes and polishes, but above all leather. Stacks of the stuff piled in sides, and shoulders and splits. Cowhide, goatskin, sheepskin. Raw hide, veg-tanned, chrome-tanned. It was an extraordinary new world.

I started by stitching a simple sleeve case for my iPad, marking evenly-space stitch holes with a dinner fork that my Jacqui never missed, and making the holes with a sharpened nail. Since then, I have progressed to handbags and cases for Jacqui and her friends, sheaths for knives, and boxes for stuff. One design I have returned to over and again is a wet-moulded leather bowl, double-skinned and hand-stitched, and spattered with dye in a Jackson Pollack style! Time spent on these projects is quiet time, days in my shed with needles, knives, the smell of beeswax and mink oil is rarely wasted, but I am still a lifetime away from the skill of the cobblers of Northampton. I reckon I have completed my first thousand hours, which has brought me to the foothills of their art; from here I can at least appreciate the towering heights of their skill.

An Artefact

One day during a Google search my eye was caught by images of an ancient and worn leather bottle that was brought up from the English Channel with the *Mary Rose*, a fighting ship that served in King Henry the Eighth's fleet before sinking off Portsmouth in 1545. When the wreck was recovered in the 1980s, one of the many artefacts recovered was the leather bottle listed by the curators as 81A2218. It is as black the mud from which it was rescued, but remarkably is still in one piece.

The ship went down during a skirmish with the Spanish fleet, the only ship sunk that day. Late in the afternoon, and in a rising breeze, she broached during close manoeuvres, and with her starboard gun ports left open in the heat of battle, the *Mary Rose* shipped water and sank instantly. Of the 500 crew and marines aboard, only 35 survived. Most were trapped beneath the netting that covered the decks, to hamper a boarding attack[34].

---

[34] - The events leading to the sinking and the sinking itself, are accurately described by author, TJ Samson in his fifth Shardlake novel, 'Heartstone'. It is grim, but compelling, reading and Samson is true to historical sources.

Technically, the design of the bottle is that of a 'flacket', or a pumpkinseed- or pear-flask, for its shape. It was made from two pieces of leather, one for the front and one for the back, moulded wet and stretched to give it volume, then it was dried and proofed to hold liquid. That the *Mary Rose* flacket survived almost completely intact for more than 500 years buried in mud and salt water, is testament to the leather and to that nameless craftsman or woman, who made it well.

What is perhaps surprising is that the flacket still holds a secret as to how exactly it was made. A debate over its construction does not exactly 'rage', but it certainly glows in a small corner of the internet inhabited by a community whose interests are in reproducing ancient artefacts by traditional methods.

The discussion about the *Mary Rose* flacket centres around whether the leather halves were wet moulded separately over wooden formers then dried, shaped, proofed, and finally stitched. Or, if the two halves were cut, then stitched together, soaked, and stretched by filling with sand or similar, and thereafter proofed.

For the craftsman, both approaches have merits. The first needs fewer processes, and allows more reproducible results, but it requires considerable effort in the shaping of male and female parts of the formers. The second approach seems to me more interesting, more organic and clearly requires only minimal tools – so that is the way I did it.

Fabrication

I began with a side of vegetable tanned leather – a sheet entirely recognisable as the side of the beast it was taken from. The leather I use comes mostly from South America, where it is scraped to a roughly even thickness, tanned in a solution made from oak bark, and then stretched across wooden frames to dry. I use the cheaper cuts, just because they're not perfect and bear the marks of lives lived by the beasts. Many have the visible mark of a brand, the scar of a wound or infected insect bite – all tell a story. For me, all add character to the final piece, and are acknowledgement of the life lost in the process.

For my bottle, I chose a 2.5-mm thick leather and traced the pattern I had drawn from the images of the *Mary Rose* flacket onto the smooth, 'grain' side of the leather. Before I cut, I always sharpen my knife. I take down my father's stone, oil it, then run the edge of the knife up the stone – remembering the movements he had me copy, and the shape of his hands holding the blade. He had flat spades for thumbs, the same thumbs I see today on my own hands, and I recall the precise motions he used to sharpen a straight blade or rounded knife.

A trained leatherworker will use a semi-circular knife to cut leather, pushing it forward through the leather. This technique allows her to see the cutting edge and exactly where the knife is going to cut. A line cut this way can be absolutely precise, staying in the very centre of the marking-out line. However, pushing the knife is a skill I have yet to master, so I draw my straight knife towards me in the conventional action, watching where it has already cut, and banking on luck to stay in the centre of the line. I go slowly and concentrate on keeping the blade cut vertical and the cut true. When brought together my two halves match up well enough.

With dividers, I work round the perimeter of each half, running one arm of the dividers around the edge, while the other arm traces a line two millimetres in from that edge. This marks the two channels that will take the stitches. Now with a pricking wheel, I mark the stitches, about 300 hundred in all. The pricking wheel is a tool of such simplicity and perfect function that it brings me joy every time I use it. It is simply a rotating cog mounted on a handle, but when rolled gently over damp leather it leaves a trace of evenly spaced marks that will guide my stitches. I follow curves and corners, round each half. There's always a chance to skip or slip, so I count all the stitches, making sure each for every stitch on one half, has its partner on the other half. One extra mark, a single diversion from the regular lines, marks the single "V" stitch, I will make, and which will identify this work as mine – my signature.

Now I lay each half on a pricking mat, an offcut from the rubber mats we laid in Jacqui's horse's stable. In the marks made by the pricking wheel, I push my awl through the leather. Bringing the halves back-to-back, I stitch them together. Two needles passing through each pair of holes in opposite directions and drawing the two sides together. This is saddle-stitch, the strongest stitch - a labour that requires patience, but one that is perfect, authentic and beautiful. The stitches done by sewing machines are quite inferior. A machine stitch is one where one thread is looped against the other leaving a point of wear that is prone to weakness. A machine stitch cannot be pulled tight without breaking the thread. Here I want a waterproof joint, so I will need very tight stitches. I saddle-stitch by hand, with thread treated with a ball of beeswax and turpentine to lubricate and protect, it is pleasingly fragrant. Like a repeating mantra, stitching is meditative, the turpentine calms me, and the rhythm cannot be rushed.

I always thread the holes, left-to-right then right-to-left, pulling the two threads taut, this is hard on the hands but gives a neat and strong stitch. On a straight line, I can manage about 5 stitches per minute, and can do half-an-hour before needing a rest. Joining the two halves of my bottle with two lines of stitching takes most of an afternoon.

Next, I take the two halves, stitched but still flat, indoors. Submerged in water that's just a little too hot for an ungloved hand, the leather fizzes as the air comes out of it. After soaking for about twenty minutes, it emerges transformed. Now it is a heavy and the colour of dark toffee, and it feels almost like jelly. I open the space in the stitching that will form the neck of the bottle and push a funnel down into the open mouth. I fill the funnel with sand and shake it down into the soft wet stomach of the bottle.

My bottle distends and grows heavy. I hang it from a loop round a beam in the roof of my workshop and continue to fill. When full, I take a copper hammer, and beat the leather to soften it until I can pour in another cup of sand. The copper hammer once belonged to the father of my previous partner. He used it to beat out the dents in his ancient cars and had once used it to make a mudguard for his Morgan three-wheeler, beating it over his own thigh, which seemed to be about the right shape. It was a long-time ago that I loved his daughter, and in a different way, loved him too. When his daughter and I split up twenty-years ago, it seemed right that I cease contact with him, but he gave me his hammer, and through it we are still connected. I use this hammer often to pound the leather without leaving the marks that an iron or steel hammer would leave through some chemical reaction that I do not quite understand.

The pounding of the leather relaxes the leather further and allows more sand to be accommodated. After an hour, the bottle hangs bloated, looking like a full, ripe fruit, hanging on its hook.

It takes most of the week to dry. I visit often to touch it. Its surface is still plump, smooth, and clammy to the touch. It feels more than just dead leather.

When dry, the leather has stiffened a little, but the shape is harmonious, full, and bulbous. I am pleased with it. My eye is good enough to see that it is not perfectly symmetrical, but it is close enough. It would have been perfect if I had chosen to mould it, but I like that this form has grown from the leather itself, defined by the stretch the material itself was able to accommodate. It wasn't forced to meet a pre-set form. This is a more organic process working with the nature of the leather itself, its weaknesses and its strength that guides the shape.

The method I chose does, however, have one inherent drawback, and I sit for a long time with my flacket in my hands, wondering how to proceed. This method of moulding requires that it is stretched wet, and so waterproofing of the leather must come after the shaping. So, I have in my hands, a flacket in name only, a good-looking ornament. Form without function, because it is unable to hold water, the job for which it was made. I contemplate trying to apply the proofing to the inside through the narrow neck, but this will be impossible. So, with a heavy heart, I slide a razor blade between the halves and cut away the stitching.

Pulling apart the two halves, they lie like avocado skins on my bench. They are once again parts, but I have access once again to the interior. I warm beeswax and pine resin then paint it onto the leather. Mixed and melted together in equal measures, the beeswax waterproofs and the resin stiffens, holding the wax permanently. After each coat, I warm the leather, taking care not to scald it, and watch the wax soak in. I apply many coats. When done, I marry the two halves together again, and use the same holes to replace the stitching as before. This time the thread drags on the resin and each stitch must be perfect, the work takes even longer. I stitch with the cork in place, pulling tight around it to ensure a good fit.

I proof the outside of the flacket and along the stitching, and then dye it. Mixing the dye with carnauba wax, which will give a lustre that will last many years. Finally, I sand and buff the raw edges of the leather and polish them with wax. I thread on a strong strap taken from a broken horse bridle and hang the flacket for a final test. It holds almost a litre of liquid, and with some remedial additions of wax is waterproof. Water left in it for more than a few minutes tastes bad, but I reckon that might pass after a couple of decades of use.

Over a couple of weeks, my leather bottle took, perhaps, 40 hours to complete. It gives me pleasure to hold it in my hands. Perhaps not an exact facsimile of the *Mary Rose* flacket but certainly a good likeness, and it has served my purpose. Absorbed by my task, I was connected to something I have loved, and was rooted in place and craft.

Antarctica explained

Throughout this book I have tried dear reader to give you some insight into Antarctica, climate change and me. Antarctica is an extraordinary place, so to help put things into context I offer a summary of how this vast uninhabited continent sits within a framework of international co-operation. Of course, for enthusiasts there is a huge amount of published information, and whether you are familiar with Antarctica or perhaps have only a passing interest, I would certainly encourage you to visit the British Antarctic Survey website. Here are just a few key points of interest.

International governance

Antarctica is remote and weather is hostile. Up to four kilometres of ice covers the bedrock. There is no native population, and it is dark for half the year. The people who do live and work there are from the 28 or so nations that operate scientific research facilities. Tens of thousands of tourists visit each summer, mostly on ships. This frozen continent is critical to understanding how our world works, and our impact upon it.

On 1 December 1959 the governments of twelve countries signed an international agreement in Washington. Seven of those had made territorial claims in Antarctica. Ratified in 1961, the Antarctic Treaty designated the continent as a natural reserve, devoted to peace and science. Today, around 56 international 'Parties' meet each year to negotiate and agree the governance framework that protects and conserves Antarctica's environment and its wildlife. The treaty is enshrined in the national law of countries that operate in Antarctica. Within the UK, the Antarctic Act (1994 and 2013) implements legally binding agreements under the Antarctic Treaty. The Foreign, Commonwealth and Development Office Polar Regions Department is responsible for the administration of British Antarctic Territory. BAS provides expert advice to Government and is part of the UK delegations at the Treaty meetings.

The Scientific Committee on Antarctic Research (SCAR) is part of the International Science Council (ISC). Created in 1958 it initiates, develops, and coordinates high quality international scientific research in the Antarctic region (including the Southern Ocean), and on the role of the Antarctic region in the Earth system. SCAR provides objective and independent scientific advice to the Antarctic Treaty Consultative Meetings as well as other organizations such as the United Nations Framework Convention on Climate Change and IPCC.

The Council of Managers of National Antarctic Programs (COMNAP) was formed in 1988 to bring together leaders with responsibility for delivering and supporting scientific research in the Antarctic Treaty Area on behalf of their respective governments. COMNAP supports the free exchange of information between nations through its expert groups including air operations, science facilitation, health and safety, environmental management, and protection.

The business of exploration

Since antiquity, exploration has been fuelled by commerce. Exploration, followed by conquest and colonisation opened access to raw materials and to new markets. The greatest exponent was probably the Roman Empire. Its thirst for trade even more than its thirst for territory sent military explorers and cartographers across Europe, Africa, and Asia. In AD 61, in a prelude to a planned invasion of Ethiopia, the emperor Nero sent a praetorian guard to find the source of the Nile. They traced the White Nile at least into South Sudan, and there are some historians who believe they reached Lake Victoria. This first in exploration was repeated and claimed as a first by the nineteenth century explorers, including Livingstone and Stanley, many of whose expeditions were funded on the promise of trade.

Even the European empires that persisted into the 20th Century, were built on the efforts of the few men, and even fewer women, who ventured into parts of the world that were considered unknown, except of course, to those who already lived there. However, in the early part of the 20th Century, the ability to map from aircraft meant that maps could be filled in without the need to trudge through jungles, across deserts, or icy wastes.

For a few decades, the caste of European men who would have been explorers were found other roles, mostly at war, but with the end of the WWII, many looked again to make their names in wild places. True 'exploration' took to space, and perhaps into the depths of the oceans, but a new breed was born, and a new business model developed.

They called themselves 'explorers', but these men were professional adventurers. For examples, think of: Thor Heyerdahl, Jacques Cousteau, Reinhold Messner, Sir Edmund Hilary and Sir Francis Chichester. Each forged successful careers by going to places beyond the reach of mere mortals, but more importantly, by undertaking feats of unimaginable endurance and skill.

To add spice and grandeur and attract publicity, many claimed firsts for their adventures by multiplying and sub-dividing records set by those that had gone before.

For example, Joshua Slocomb completed the first solo global circumnavigation under sail in the late-1800s, although he stopped several times to resupply and rest. On this basis, Sir Francis Chichester's historic journey in 1966-67 was conceived as the first non-stop solo global circumnavigation. In the event, Chichester was forced to stop to effect repairs, and so on his return, rather than 'non-stop', his voyage was hailed variously as the 'fastest circumnavigation in a small boat' and 'the longest single passage in a small boat' etc. The first non-stop, solo circumnavigation was claimed by Sir Robin Knox-Johnson in 1969. In 1971, Chay Blyth completed the journey, solo non-stop but in the 'wrong' direction against the prevailing winds. And so on, and so forth. Nowadays, with recent advances in sailing and in navigation technologies available, the fashion is for the fastest times and the youngest sailors.

These professional adventurers developed a business model that is not much different to that used for a century and more by explorers and is still used today. They spent years accumulating the finance for each venture from commercial sponsorship, and on return, maintained an income stream through book, film and broadcast royalties and appearance fees. It was not a very big business, and was certainly a precarious one, but it was business, nonetheless. And they soon realised that what sold was danger, hardship, and heroism.

While many of the new breed crossed the oceans in increasingly crazy boats and climbed increasingly impossible routes on the world's most treacherous mountains, a few made the polar ice their playground, seeking out new firsts in polar travel.

In 1993, Ranulph Fiennes and Mike Stroud attempted the first unsupported crossing of Antarctica. Running out of time and food, the two abandoned their plan before crossing the Ross Ice Shelf and were evacuated via South Pole and Rothera Station. Redefining success, they subsequently claimed the first unsupported 'coast-to-coast' crossing of Antarctica, meaning they had missed out the sections they had planned over the ice shelves, and cut out about 1500 km of trudge, but crossed the ice sheet where it rests on bedrock.

Børge Ousland, a Norwegian, finished the first real solo Antarctic crossing, floating-edge-to-floating-edge on 18 January 1997. His journey was 'unsupported' in that he did not use pre-laid depots, but 'assisted' in that he employed a kite to help in propulsion.

However, even this term 'unsupported' is divisive. Adventurers claim to be 'unsupported', in the sense that they carry everything they need from start-to-finish, the reality is most are 'supported' every step of the way. Their location, and sometimes even their bodily functions are monitored daily, by teams who are available instantly over satellite phones. They may receive daily advice and weather forecasts. Rescue, should it be needed, can be summoned at any time. The security that this provides, allows longer and harder routes to be planned and executed by less and less experienced people. It is hard to imagine that, in 1909, Shackleton would have turned back just 112 miles (180 km) short of the South Pole if he had known that an airlift or food drop could have been summoned during his return journey.

American Colin O'Brady became the first person to complete a solo, unassisted, and unsupported crossing of Antarctica in 2018. He completed 1,700 km in just over 54 days, including only 15 days to cover the last 600 km from the South Pole to the edge of the Ross Ice Shelf. However, in this last section he skied along a 'road' that had been surveyed and groomed by USAP during their resupply of their station at South Pole. O'Brady could ski fast on this unnaturally smooth surface, and crack on without wasting time navigating, or looking out for crevasses.

The last 'great' expedition attempted in Antarctica, was both so impossibly arduous and insanely difficult, that a team of the hardest, most committed, and toughest ever assembled did not achieve even a tenth of the route planned. In 2012, Sir Ranulph Fiennes' failed in his attempt to cross 2000 km of the high East Antarctic plateau during the dark of the Antarctic winter. Described at the time by *The Guardian* newspaper as an 'appalling challenge', it was bizarre in its concept. Fiennes called his expedition 'The Coldest Journey'.

Aged 68, Fiennes and a companion would ski in front of an 'ice-train', consisting of two 20-tonne Caterpillar D6Ns snow-equipped tractors, and various living cabooses, storage, and fuel sledges, supported by a 4-man expedition team. In areas of uncertain ground, Ranulph could continue to ski, but lest he should fall into a crevasse he would be tethered to the tractor.

Sadly, Fiennes was forced to withdraw from the team just before the trek started in earnest due to a nasty case of frostbite during his training on the ice. He was evacuated to South Africa. The remaining team, led by the traverse manager, Brian Newham, nonetheless, started out to complete the journey without the skiing. Shortly after mid-winter, Newham was forced to abandon the effort after 300 km, trapped in the dark in an extensive crevasse field. In the dark, the team was neither able to move forward or backward, and so had no option but to wait until the light returned months later, before retracing their steps. Although, the expedition blogs make an interesting archive[35], I have found no comprehensive published account of the expedition, and the question for me still hangs —Why?

Each of the individuals who appear in this cabaret of polar adventure has undoubtedly been extraordinary in the true sense. Each stepped into their skis and harnessed themselves to their pulk sledges piled high with high-tech freeze-dried food, and scientifically calculated quantities of fuel for their space-age stoves. All

---

[35] - https://www.thecoldestjourney.org/blog/

faced Herculean tasks, that most of us, me included, could not imagine. All are individuals of extraordinary strength, stamina and drive, and to-date they have criss-crossed the continent so many times that there is barely an unused route left to try, and fewer ways to make it harder. But, 'unassisted', 'unsupported', 'solo', in the rough or on a road, towed by a kite or tied to the front of a tractor; for me it has all got a bit too weird!

Notwithstanding those mentioned, there are polar stories that I am sorry are not better known, because I personally find them truly inspirational or cautionary. For one reason or another, these are polar stories seldom told.

Will Steger - Great expedition, poorly recounted

One of the most extraordinary polar journeys attempted in recent decades was the 1989/90 'International Trans-Antarctica Expedition' led by the American, Will Steger. This expedition was historic in that it was the last, and certainly the most ambitious Antarctic journey to use dog teams[36], and it followed the longest possible route across Antarctica, from the tip of the Antarctic Peninsula to the far side of East Antarctica, via South Pole.

For the so-called, 'purists', this expedition was, or course, tainted by being 'supported', because aircraft were used to resupply the team with food and fuel along the way. For me that isn't such a big deal; the distances Steger's teams had to cover were enormous and he would have run into other problems if he had suggested a more traditional approach. For example: the approach used by Amundsen, who set off for the South Pole with 52 dogs then progressively shot the weaker dogs and fed them to the stronger dogs, and humans, so that he returned with only 11.

---

[36] - Sledge dogs (huskies) were vital to many nations' activities in Antarctica until the 1980s. British Antarctic Survey sent dog-teams on summer expeditions around the Antarctic Peninsula and then only gradually replaced them with skidoos. During my first three seasons in Antarctica, there were still three dog teams living outside at Rothera Station. Our dogs were much loved by almost everyone on the station, and they provided unending, non-judgmental support to anyone in distress.
However, the Protocol on Environmental Protection to the Antarctic Treaty, which was signed in 1991, forbade further use of dogs in Antarctica. In part, this resulted from a fear that canine distemper might be transferred into the native populations of seals. The protocol stated simply, "Dogs shall not be introduced onto land or ice shelves and dogs currently in those areas shall be removed by April 1 1994". In 1994, the 14 remaining BAS dogs were removed from Rothera Station. After a 5-hour flight to the Falkland Islands, where the dogs spent a few weeks adapting to the warmer climate and new surroundings, they were flown to the UK, and thence to Quebec in Canada. Of the 13 dogs who arrived in Quebec, 5 died within the first year due to infection and disease. A sad end to a long partnership of service and friendship.

Steger's journey took seven months and was unusual in being completed largely without issue or serious injury to human or many of the dogs. The team was truly international: Steger as leader, a French doctor, an experienced Russian Antarctic scientist; a dog handler from Japan; a Chinese glaciologist, and a Brit, Geoff Somers, a BAS veteran who served as the group's navigator. They set off in good time, got on with one another to the extent that they needed, and their goals were achieved without much issue.

In fact the International Trans-Antarctica Expedition was so well run that even for me the official account[37] made a dull read. One masterful, if discouraging, review[38] of Steger's book described it as a 'routine account of a ground-breaking expedition', one in which the authors 'try hard to make their tale a gripper'. However, the reviewer correctly identified the narrative problem in the expedition itself, 'which was masterfully organized and proceeded without major mishaps. The reviewer summarises the narrative, 'It snows a lot. Everyone gets home safely', and he recommends readers looking for thrills to 'try Shackleton or Scott'.

I would add only that, owing to presence of Qin Dahe, an eminent Chinese glaciologist and eventual co-chair of IPCC Working Group I, the expedition was not quite unique, but certainly noteworthy, in that even the scientific data collected enroute were of sufficient quality and novelty to be published in the scientific peer-review literature.

That is how it should be, but of course, expertise and efficiency do not sell books, and do not fund the next big trip. Drama sells books; frostbite sells, falling down crevasses and running out of food and misery in all its forms sell. The polar narratives we buy are those of extreme privation, and the astounding assault by the forces of nature, and where expeditions like Steger's are told without hyperbole then they tend to be forgotten. I hope that for Steger and his team, the fact that their story is rarely told is a triumph rather than a disappointment, even if there are only a few of us who recognise their skill.

Augustine Courtauld – a quiet man

A similar fate befell my personal favourite seldom-told tale of polar exploration. This one happened shortly after the end of the heroic era, not in Antarctica but in Greenland. Unusually, in the polar canon, it is an interior rather than exterior story.

---

37 - 'Crossing Antarctica', by Will Steger, and Jon Bowermaster, published by Alfred A. Knopf 1992

38 - https://www.kirkusreviews.com/book-reviews/will-steger/crossing-antarctica/

It concerns one Augustine Courtauld, a cousin of the wealthy industrialist Samuel Courtauld, and a rather well-to-do and impossibly good-looking lad. He had previously been a stockbroker, although not a very good one, who reputedly lost his employers a great deal of money through a misplaced decimal point.

In 1930, Augustine joined Gino Watkins in on an expedition to map out a northern air route from Europe to North America over Greenland. While the story of this expedition includes many errors and mishaps, the account of Augustine's survival alone on the ice through the Greenland winter is a fascinating one.

Having trekked high up onto the Greenland plateau, and with supplies running too low to support the planned two-man occupation, Courtauld already suffering from cold, volunteered to over-winter alone in a remote high-altitude station. For the next several months he would make the planned meteorological records every three hours waiting out the winter until his colleagues returned to pick him up. The issue was that his 'station' consisted of just one tent and two igloos, each protected by a wall built from snow blocks. He had no radio and no means to leave.

Within a few weeks, Courtauld's residence was all but buried in snow; probably because those protective walls attracted drifting snow around his camp. He managed to dig and maintain a system of tunnels between the tent and igloos, and to the outside, and managed to continue making the meteorological observations for many weeks. However, eventually the tunnels were overcome by the sheer weight and persistence of the snow drift. He tried to maintain tunnels up and out, but the tailings from his digging rapidly filled his living quarters – soon Courtauld was imprisoned below the ice. To begin with he was comfortable and warmed by a paraffin stove. He filled his time reading by candlelight; but as his food, and more importantly his fuel, ran low he was reduced to lying alone in the cold and dark, simply waiting for his colleagues to return and locate him.

Above ground, his colleagues returned with the light but struggled to find the now buried station. For days they searched. Having all but given up hope, the surface party eventually located the tops of a couple of meteorological instruments, the tattered Union Flag, and a tin chimney that remained visible poking out above the snow surface, although this had been cold for days. They called down the chimney and were astounded when Courtauld replied.

By the time Courtauld was rescued he had spent five months alone, his food and fuel were completely exhausted. For several days, he had laid alone in the dark and cold waiting for rescue or death.

The physical hardship Courtauld endured was impressive, but for me, the real story is about his inner experience, and how much, or indeed how little, it impacted his mental health.

There was surely some bravado involved when he later reported,

> *"One cannot be bored living an entirely novel life under such interesting conditions. My physical and mental condition, the weather, speculation about the work of the expedition and the doings of friends at home were subjects that fully occupied my mind. I never had the slightest doubt with regard to my relief, though I fully realized that it might be delayed."*

Nevertheless, such fortitude and inner resilience is astounding. After a haircut he slotted straight back into the expedition team, taking part in several new trips; he even led some notable boat trips to map the Greenland coast.

To read Courtauld's letters, he was far from the devil-may-care, macho, boy's own hero. Rather he seems a rather gentle and sensitive soul; indeed, it was, perhaps, his deep and passionate love for his future wife, Mollie, that saw him through those desperate months. In the years that followed, Augustine and Molly's long and happy marriage seems to indicate an easy-going character.

Perhaps, resilience was not inbred but trained into Courtauld. The English boarding school to which he was sent at an early age, was clearly one where bullying, punishment and probably injustice were institutionalised, even weaponised, against young boys. And in that establishment, he was not on the top of the pile. Some were probably broken, but that early mistreatment seemed to strengthen Courtauld and increased his capacity to endure or accept.

Returning to England, Augustine himself refused to dramatize the events on the Greenland ice sheet, and so again – as with Steger – it was the apparent nonchalance with which his story was told that condemned his exploits to a footnote in the history of polar exploration.

Today, in a world where an individual's mental health can reportedly be 'damaged' by any number of life's upsets, resilience is for me something to strive for and something to admire. I admire that resilience that allowed Courtauld to walk into the light from what must have appeared to be his icy grave. Allowed him to take a haircut and shave, smoke a cigar and regain his equilibrium and carry on. Augustine Courtauld, I for one salute you for your fortitude and patience, and for your restraint in the telling of your story.

A Southern Ocean emergency

During a typical austral summer British Antarctic Survey works across a vast area of the South Atlantic and Antarctica, and occasionally things do go wrong. In the best planned and managed ventures, people do get into trouble, and organisations like BAS take the lead role because of their mission as national operators and because they alone have resources to deploy.

During my time as Director of Science for BAS, I was involved in one such incident. I took part in BAS's response to a medical emergency in the South Atlantic. The case involved the evacuation of a critically ill individual from a subantarctic island[39].

My role for which I had been trained, was as Leader of the Gold Team, one of the key roles set out in BAS's Incident Response Plan. This response plan mirrors those developed by emergency services in the UK and elsewhere. The underpinning idea is that no matter the exact nature of the emergency the organisation's response is handled in the same way. Individuals slot into teams whose purpose is well-defined, as are the roles for which they are trained.

In short, the Bronze Team works at the site of the emergency doing everything required to preserve life and ensure swift conclusion of the emergency. They are supported by the Silver Team back at BAS HQ in Cambridge, and a BAS Medical Unit made up of a specialist team of doctors and clinicians located in a hospital in Plymouth. The Gold Team provides leadership and support to the two other teams, drafts press releases and maintains high-level communication with NERC, UKRI and Government. Gold also establishes and maintains dialogue with families of individuals affected by incidents. This allows the operational teams to focus on emergency actions.

For almost two weeks, our response teams dropped all other tasks and worked day-and-night to help our poorly colleague. Multiple options to repatriate the patient to a hospital were explored. Medical experts who could provide life-saving specialist expertise not available on-site were mobilised. Throughout the effort, I was amazed at the expertise brought to bear, and effort required and expended to secure a satisfactory outcome.

---

[39] - In this case, preservation of the medical confidentiality of the individual concerned is still important and a continued responsibility of those of us involved. I will tell the story in summary to respect that confidentiality.

Two BAS ships, and several tourist vessels in the area were on standby to assist. Several important science programmes were put on hold, and others were cancelled completely. Later military vessels and their personnel based in the Falkland Islands were engaged. Although an airborne evacuation was not possible, several aircraft were held on standby in the Falklands to provide onward transport north. And finally, a specialist military team that had expertise in managing critically ill patients during repatriation flights was deployed from the UK to wait in the Falklands Islands.

One of my roles was liaison with the patient's partner, who was understandably extremely shaken. I did my best to ensure they were kept informed about all developments in twice-daily phone calls.

For a couple of days, the lack of medical oxygen for the patient was a serious problem. Tourist ships in the area donated their supplies. Oxygen generators, which extract oxygen from air, were transferred from a nearby BAS station onto the RRS *James Clark Ross*, which would undertake the evacuation. Oxygen stocks were too meagre to risk putting our casualty onto the ship for the minimum five-day crossing back to the Falkland Islands. Such concerns are not routine in well-stocked first-world medical facilities, but in the South Atlantic, they can mean the difference between success and failure – in our case, life, or death. Thankfully, the ship's engineer suggested using the ship's supply of welding oxygen. It took a matter of hours to determine that this oxygen was sufficiently pure, and that our ship's engineer could rig up the required valves to connect welding and medical systems together. With this break-through the casualty was immediately consigned to the ship and set off.

Throughout the voyage, the patient's condition was already serious, but on the approach to the Falkland Islands it began to deteriorate. Once the ship came within military helicopter range of the Falkland Island, the decision was made to uplift the patient from the ship and fly them direct to Mount Pleasant where a long-range aircraft would be waiting on the tarmac to take them north. This was bold because winching the patient from the deck of a ship in the Southern Ocean during rough weather was risky.

At the end of two weeks our casualty was safe in an excellent hospital in South America; reunited with their partner who flew out from the UK with a representative from their university; BAS's emergency teams were stood down. In all, a couple of hundred people across multiple organisations were involved.

I tell this story respecting the confidentially of the individual concerned. The rescue of people in and around Antarctica – whether they be there on scientific business, private expeditions, or tourist excursions – requires huge effort, and can, on occasion cost a great deal in human and cash terms. All nations and organisations operating in Antarctica will go a long way beyond their normal operating parameters to rescue people in need.

It is vital that all expeditions, nationally or privately supported, do all they can to avoid accidents that might require outside assistance.

As a postscript, it is worth noting the safety record of my organisation, British Antarctic Survey. From 1948–1982, while operating several year-round stations in Antarctica as well as several ships and aircraft, BAS lost a total of 28 souls in accidents or due to ill-heath whilst deployed. Advances made in risk management, enhanced safety procedures and improved support systems reduced this number to zero until the tragic loss of Kirsty Brown who was drowned by a leopard seal while snorkelling at Rothera in 2003. It is impossible to prevent all accidents in the challenging and remote conditions in which BAS operates, but devotion to safety is evident in everything BAS does.

The Aurora Programme – when things do go wrong!

Another Antarctic rarely told story is that of the Aurora Programme of 1991/92. This expedition was led by a truly charismatic Norwegian, Dr Monica Kristensen. I met Monica a couple of times, she glowed with a particularly Scandinavian type of vitality and capability. She could be charming, but behind that charm was an unstoppable determination.

Monica was a glaciologist and climate scientist of some renown. In the early-1980s, she had earned her PhD at the University of Cambridge's Scott Polar Research Institute, researching the interactions of icebergs with ocean waves. She also had a personal obsession with Roald Amundsen. In 1987, she led an expedition attempting to recreate Amundsen's South Pole trek with British glaciologist Neil McIntyre and two Norwegian dog-handlers. Although unsuccessful, this expedition was widely applauded, and set the stage for something far more ambitious.

The inspiration for the Aurora Programme was Monica's ambition to find, recover and return to Norway the tent which Roald Amundsen had left at South Pole on 14th December 1911. She planned to present Amundsen's tent as the centrepiece of the 1994 Winter Olympics in Lillehammer. On the back of this high-profile, and undisputedly popularist objective, she built a sophisticated science programme, largely aimed at providing calibration and verification data for the soon to be launched European Space Agency satellite ERS-1. With several angles to pitch, Monica attracted considerable financial support and sponsorship for her expedition, including from Statoil in Norway, and she acquired prominent scientific partners including the Climate Physics Group at the Mullard Space Science Laboratory (MSSL)[40]. At the Royal Society in London, and later in Oslo, Monica launched the Aurora Programme, named after her expedition ship, the MV *Aurora*.

Given the breadth of her objectives it might have been argued that the programme was already over-ambitious, but she was persuaded to trade simplicity for wider Nordic support. Field-parties from Norsk Polar Institute and University of Stockholm were added to the roster. In addition, the time constraint of reaching Lillehammer with its prize, meant that the programme relied in parts on inexperienced team members.

Nonetheless, the Aurora Programme was scientifically bold and significant. So, despite some intense but hushed lobbying against it by the polar establishment, its momentum was unstoppable, and the MV *Aurora*, complete with a helicopter, set sail for the Antarctic in Autumn 1991. However, having languished in Montevideo waiting for the delivery of cargo and for striking ship workers to load fuel and supplies, the expedition reached the edge of the Antarctic sea-ice in late-December well behind schedule.

Over Christmas, *Aurora* was stuck fast in the sea ice, but on New Year's Eve it came within helicopter range of the BAS's Halley Station. A friend of mine who was on the expedition recounted the events that followed, '…we got the [Aurora's] cook to bake a pizza and flew into Halley, knocked on the door, and asked who had ordered it… BAS could not formally welcome us. However, the station staff were delighted to have visitors, and several of them were flown back to visit the Aurora'[41].

In mid-January, the *Aurora* reached the edge of the ice shelf, where it was joined by a ski-equipped, Twin Otter chartered from Greenland Air. Moored against the ice cargo and scientific equipment was unloaded, and moved to an inland site, where a collection of prefabricated huts was installed inland to provide a base for activities. At this point, deployment of the field parties could begin; however, the sun was already

---

[40] - At the time the Climate Physics group at MSSL was led by Chris Rapley, who would later be appointed as Director of BAS. Rapley welcomed the opportunity to partner with Kristensen in the Aurora Programme and sent several of his staff on the expedition.

[41] - It does not surprise me that the Aurora team got cold official reception at Halley Station. The official UK policy, dictated by the Foreign and Commonwealth Office, was that, except in emergencies, the UK represented by BAS would not provide 'support' to any expeditions to Antarctica that were not under the auspices of a national Antarctic operator. By this definition, the Aurora expedition was a private expedition. For the same reason, Monica's team got a similar reception at the US Amundsen-Scott Station at South Pole.

dipping towards the horizon, and it was very late in the season to be deploying remote field parties this far south.

My friend was deployed to a southerly site on the Filchner Ice Shelf in a three-man party. Only one of this party had any Antarctic experience, a single brief airborne season, and they took a meagre supply of safety equipment. However, this team were under considerable pressure to complete their scientific projects. Some of his stories made my toes curl. On one occasion he found himself, 'alone, at 3am, 40 km from camp, at -25°C, and with no radio contact'. To be fair, at this point he decided that turning his skidoo off to avoid interference with his instruments would be foolhardy, and suddenly realising the risks returned to camp *tout de suite*, but later he wrote, '… it all turned out fine, but yeah, it was bloody stupid. But we had come all that way, and I wanted my data'.

Meanwhile, Kristensen herself was heading to the South Pole on her quest to find Amundsen's tent. She arrived in the Greenland Air Twin Otter on 16th of February, with Heinrich Eggenfellner, and a new-fangled ground-penetrating radar (GPR) system that would allow them to locate Amundsen's tent buried beneath almost one hundred years' accumulation of snow. They were only able to spend a few hours at South Pole but the folk at Scott-Amundsen Station provided them with a vehicle. Having surveyed the area with the GPR, Monica and Heinrich, pitched a tent and planted a flag at the site they believed to be the location of Amundsen's buried tent.

The Aurora Programme suffered several other setbacks. Most significant was damage to the Greenland Air Twin Otter which landed at a poorly surveyed airstrip that turned out to have crevasses on it. Having dropped a ski into one of these crevasses, the aircraft had to return to Canada for repairs. During the several weeks that journey took, the field parties were effectively 'stranded' out on the ice.

With the Twin Otter back in Antarctica, Kristensen eventually called the end of the season, her field parties were gathered in, and the programme regrouped on the *Aurora* and headed north. They returned to Europe where Kristensen received a critical, if not actually hostile reception, and various accusations, some financial, some environmental, and some with regards to safety were made in private and public. It was clear that support for Aurora to return to Antarctica the following year had collapsed.

Kristensen regrouped around her core ambition to retrieve Amundsen's tent, and so in late-1992 only a skeleton team returned on an ANI flight to Patriot Hills. This team undertook a traverse to South Pole, once again with the intention of retrieving Amundsen's tent in time for the Lillehammer Olympics. However, this final effort ended in tragedy and controversy.

At 8:15 p.m. on 28th December an urgent request for assistance was received at McMurdo Station. In a matter of hours, USAP mobilized a seven-person US/New Zealand search and rescue (SAR) team and put them on an LC-130 aircraft to South Pole.

According to the request, an accident occurred to a four-person group enroute overland to the pole, as they were crossing a heavily crevassed region, somewhere in the vicinity of Support Force Glacier. The injured party had fallen into an unseen crevasse and had been trapped there for about 30 hours.

At South Pole, four of the SAR team transferred to a Twin Otter and were deployed to the accident site. Having surveyed the location from the air and confirmed it was surrounded by crevasses, the Twin Otter landed a few kilometres from the accident site around 7:40 a.m. on 29th December, less than 12 hours after the initial call for help. From the air, the camp was an eerie sight. The SAR team could see numerous crevasses, some open and others hidden under snow bridges. They could also see tracks of four snowmobiles crossing the crevasse field. The Norwegian camp was in sight, but no signs of life were evident. About 90 m from the camp, the team saw a crevasse with ropes going down into it.

Having landed safely, the Twin Otter pilots realised that their aircraft was now surrounded by yet more crevasses. They decided that they would not return to that site again, and so any hope of ferrying the remaining SAR team from South Pole were shelved. The reduced team determined to affect the rescue on foot, and so roped together and with minimal medical and rescue gear they set off to make the 3 km journey to the Norwegian camp. The journey took four hours, during which the team fell in crevasses many times.

On arrival at the camp, they found the first casualty, who had driven his skidoo into a crevasse and reportedly fallen more than 70 meters. While he was being treated for concussion and several cracked ribs, the rest of the rescue team made radio contact with South Pole Station. During which, a second casualty had fallen through a hole in a snow bridge.

With his rope anchored to a skidoo, the SAR team leader Steve Dunbar began his descent into the crevasse. As he descended, the temperature fell, and the gap between the walls narrowed to only 20 cm. Eventually, Dunbar saw the second victim's arm protruding through the snow. It showed no discernible sign of life.

Unable to reach the body, Dunbar determined that it would be impossible to extract the body, and he returned to the surface alone. The victim was 36-year-old, Captain Jostein Helgestad of the Norwegian Army.

Returning to the Twin Otter, with the three survivors, the SAR team was forced to abandon all but the most essential survival equipment so that the Twin Otter could take off through the crevasse field in as short a distance as possible. They eventually returned to South Pole Station in the early hours of 30th December.

Analysis of the accident and response was protracted. NSF/USAP reiterated their policy that they would not assist private expeditions, except as in this case, in response to humanitarian requests for assistance. The SAR team reported that the Norwegians were unprepared to navigate the crevasse fields that they had encountered, and that neither their training nor experience was up to the demands of the environment. Consequently, it was concluded that the Norwegians had put not only themselves but also the USAP/New Zealand personnel in danger and diverted two aircraft from their duties.

While it should be noted that Monica Kristensen was not part of the traverse team, nonetheless she was leader of the Aurora Programme. Today, she would be held to account for the safety protocols in place. In some present jurisdictions, she might well have faced legal proceedings[42].

It would be harsh to use hindsight to damn completely the Aurora Programme, or indeed, Monica Kristensen's leadership of it. Despite sailing too close to the wind, in its first season the expedition required little external assistance, and everybody came home safely. The expedition also produced some decent scientific outcomes, and in a small way it helped in making the ERS-1 satellite the undoubted scientific success it was[43]. However, it was clear that risks were taken, and had things gone wrong, there was a strong likelihood that the programme would need to call for assistance from outside sources. The scope of the project spread its teams over vast areas and put inexperienced people out on a limb, pushing them beyond their comfort zones to achieve their own and the programme's goals. They were clearly exposed and at risk that even minor mishaps would spiral into major events. The fatal traverse of 1992/93 may have

---

[42] - For example, the Australian Antarctic Division together with their contractor faced corporate criminal proceedings over an accident in 2015 in which a helicopter pilot landed in a crevasse field and lost his life. An initial conviction of AAD was eventually reversed on appeal, but the process took four years and must have been awful for those involved. https://www.nortonrosefulbright.com/en/knowledge/publications/b648a29d/appeal-decision-handed-down-in-may-v-helicopter-resources-pty-ltd

[43] - See, for example: Ridley & Bamber (1995) Journal of Electromagnetic Waves and Applications, 9:3, 355-371.

suffered some bad luck, but this was accompanied by poor decisions that might have been avoided by a more conservative approach to risk, and a greater focus on safety.

In the event, the refusal of national operators (including, NSF/USAP and BAS) to engage with the Aurora Programme was, in my opinion, the right one. Riding a wave of support from multiple stakeholders, each with their own ambitions, it would have been impossible for Monika to scale back the goals of the Aurora programme to fit its limited resources. The consequence for BAS or USAP of providing support would have been a significant impact on other activities that year. Of course, USAP was absolutely the right to undertake the rescue, and their team is to be commended for their professional but nonetheless heroic actions in preventing further loss of life. However, it must not be overlooked that the SAR team had little option but to put themselves in harm's way, and it is easy to imagine some chilling outcomes in which they themselves could also have become victims.

Behind those treks to the pole

Every year our newspapers and TV news feature articles on record-chasing crossings of Antarctica. These ventures are reported in breathless prose that inevitably compares them with the journeys of the heroic age: Scott, Amundsen, Shackleton. The media may be culpable, but in my mind these modern journeys fall short in such comparisons, not just through the availability of modern equipment and technology.

Almost all these modern crossings begin with a direct flight from Punta Arenas, Chile to the commercial airstrip and camp run by the tourist company ALE at Union Glacier. After a period of preparation and training, the 'crossings' themselves begin with a short aircraft hop from the Union Glacier down to Hercules Inlet. This marks the nearest ice shelf to South Pole and so is argued to be the edge of the continental ice sheet. Whether we accept this definition or not, this point is some 700 km from the edge of the Ronne Ice Shelf, which is the true edge of the continent.

From Hercules Inlet our modern adventurers proceed inland via a series of GPS waypoints along a well-mapped and relatively safe route towards 'tea and medals', and a comfortable bed in ALE's camp at South Pole.

These expeditions are extraordinarily self-indulgent and their leaders, notoriously self-publicising. But that is the nature of their business. They are selling stories to the media, to sponsors, or increasingly commonly simply funding their trips from their personal wealth. Their tales are littered with hyperbole and the language of competition and battle. The closer to death the better. Their fitness, strength and determination are beyond doubt, but their motivation is harder to grasp. Many do support good charitable causes and claim to be making important, even significant, scientific observations, but there are easier ways to contribute to charity, and few of their 'scientific' observations add anything to the sum of human knowledge. Others seek to raise awareness of the issues that face those places, but only add to the degradation.

To visit Antarctica, or any wild place, and spend each day there focussed only on the task in hand, and the hardship it causes you, is a sad way to spend such an opportunity. To spend those precious moments between toil and sleep, posting blogs and doing interviews is surely missing the point. Antarctica is a place of extraordinary beauty and wonder and for those who take the time, Antarctica is still a place for quiet wonder, and personal growth.

Our impact on Antarctica

I was lucky to begin my Antarctic career when I did. I came at the end of an era that began in the International Geophysical Year (1958) and persisted, perhaps until the late-1980s. By this time most of the basic 'exploration' of the continent had been completed – the larger mountains and glaciers were named, the coast mapped, and the big white spaces on the map were known to be big white spaces on the ice.

A few cruise ships provided tourists with access to many of the most beautiful and stunning sights along the Antarctic Peninsula, but the interior of the continent, where I did most of my work, was the sole preserve of scientists. So, although we were few, we scientists largely had Antarctica to ourselves. We had free rein to drill our holes and stick in our flags without fear of anyone tampering with our experiments or criticising any mess we left behind.

While I followed BAS strict adherence to international requirements and spirit of the Antarctic Treaty, my impact on Antarctica, while minor, was not zero.

In the early years, on at least one occasion, I left a car battery out on the ice because our pilots feared the acid it contained could leak and attack the aluminium frame of their aircraft. I argued with the pilot, but he would not transport a battery, in what he judged to be inadequate packing.

I personally deposited human waste, rather randomly in individual holes across the ice and occasionally allowed small fuel and oil spills to soak into the snow. None of these practices concerned us at the time as everything was going to be buried by snowfall in a matter of weeks. And the sheer scale of the Antarctic landscape seemed too vast to contaminate in any meaningful sense. However, that does not make it right, and I am pleased to say none of that behaviour is tolerated in any modern Antarctic science programme. General and hazardous waste is sorted and shipped out of Antarctica for recycling. Poo is bagged and returned to stations for proper disposal.

In October 1991 the Madrid Protocol strengthened environmental protection within the Antarctic Treaty. The Protocol came into force in 1998 and lays down principles applicable to human activities in Antarctica. It prohibits all activities relating to Antarctic mineral resources, except for scientific research. Until 2048, the Protocol can only be modified by unanimous agreement of all Consultative Parties to the Antarctic Treaty.

By and large, the Protocol performs well and is largely adhered to by the nations active in Antarctica, and by commercial tourist companies. Indeed, I would go so far as to say, that it was the tour companies that led in some areas in defining and enacting best-practice. The procedures used by members of the International Association of Antarctic Tour Operators (IAATO), and the prominence that they give them in preparing their guests, led to the introduction of biosecurity measures in national programmes. However, considerable threats remain, and vulnerability is likely to increase over time, particularly on the Antarctic Peninsula, which is the focus of most of the tourist visits.

Rapid and persistent warming has been consistently measured at the meteorological stations on the Antarctica Peninsula. Not only are air temperatures rising, but the length of the summer has increased. There is an increase in the frequency of rainfall and icing events, and reductions in sea-ice extent. Not only are these changes putting existing plant and animal communities under stress, the increase in area of ground that is no longer covered by year-round by snow is increasing the opportunity for non-native species to colonise. It is the increase in the human traffic to such sites that provides a potential vector for non-native species. National operators as well as tour companies look out for seeds and soil trapped in boots or clothing.

These practical biosecurity measures help minimise the potential for accidental transport of species into the area, but nevertheless these factors are something of a time-bomb for the Antarctic Peninsula. It is hard to imagine that many decades will pass without some new species gaining a foot hold, whether it arrives on a boot, ship, aircraft, or bird. This may not be so bad if it is limited to the occasional plant that can be eradicated of at least managed. We have already seen rats and mice colonise some subantarctic islands with devastating impacts for the existing marine life (e.g., penguins and seals) and seabirds. In recent years rat eradication programmes on these islands have been successful in enabling the recovery of affected species.

I have other concerns about remnants of introduced DNA disrupting studies into the record of plants and animals that once lived on Antarctica. Environmental DNA (eDNA) is used by scientists to uncover secrets of human and biological history. eDNA persists in the environment long after plants and animals have gone. Advances in the sensitivity of DNA analysis enables scientists to detect what lived in a landscape long after the event. This is especially important for working out how and when species colonised and survived in Antarctica. For example, how some species survived periods of ice advance hiding in a few tiny ice-free 'refugia', such as high mountains and volcanic hotspots.

It is a fair bet that my personal DNA is still present on rock outcrops across Antarctica, quite possibly alongside some more DNA from the sardines, ham, and tomatoes that were in my discarded sandwiches. These may prove to be confusing for future scientists.

To protect this emerging science, and for the obvious safety benefits, I believe that there is now a strong argument that every person venturing beyond the stations in Antarctica should always carry with them a tracking device which would provide a permanent record of their movements. Such devices exist on the open market, and I for one would have been happy to carry one. Perhaps, Antarctica is one place where an individual's freedom from surveillance must be traded for access to one of the world's last pristine environments.

A note on diversity

My first job at BAS was advertised in the classified section of *New Scientist*. As was BAS policy at the time - and remember this was almost 40 years ago - only 'unmarried men between the ages of 21 and 35' could apply. Of course, this fitted me just fine, and I am sure the exclusion of women and older candidates benefitted me.

Sir Vivian 'Bunny' Fuchs was the Director of Falklands Island Dependency Survey (FIDS) until it was renamed British Antarctic Survey in 1962. He continued as Director of BAS until 1973. He was old school and opposed to the presence of women in Antarctica[44]. It was said that he was also mortally offended by the idea of any same-sex funny business in Antarctica. Sir Richard Laws who succeeded Fuchs as Director of BAS was more progressive, but only up to a point. The organisation I joined in 1985 was still in their image almost exclusively male, white, able-bodied, and apparently entirely heterosexual - but times were changing.

Life in Antarctica in the 1980s was by modern standards rough and laddish. I remember an impromptu football match at Rothera Station held out in the snow with opposing teams going at one another with gusto. The match left a queue of damaged players in the doctor's surgery, and visible hobbling around the station for days. Many of us enjoyed this culture and there was strong support to keep it all-male.

However, beneath the surface was an undercurrent of low-level bullying which passed as banter. Frequently, this resulted in physical fights, particularly in the bar. Today, we would consider the atmosphere as toxic, and for a few who did a 'full tour' of 30 months at our stations it must have been tough.

Pranks at the expense of others were commonplace. One evening at Rothera I was drawn into the doorway of the radio shack, which was already full. We listened in silence to the evening 'sched' with the field parties. At the end one of the field assistants was due to receive a personal message. To allow privacy all but the radio-operator should have withdrawn, but on this night, no one left. There was a rumour that there was to be some fun.

---

[44] - Royal Soc, 2001, https://royalsocietypublishing.org/doi/pdf/10.1098/rsbm.2001.0012

Over the next few minutes, the victim was told that a young lady from Rio de Janeiro had contacted BAS to enquire about his health and location, and to inform him that she was soon to give birth to his child. Poor radio conditions added to the drama, and the individual in question was left sitting in his tent hundreds of miles away, unable to do more than contemplate his future. The deceit was only revealed 24 hours later. I cannot imagine what he went through.

On the dining room wall at Rothera Station were hooks and tags hung on two boards. On the first board there was a hook for each bunk in the shared bedrooms (often referred to as pit rooms). Individual name tags were hung on the hooks to show who was allocated to each bunk. In the event of a fire or emergency roll call, the board allowed everyone on the station to be accounted for simply and quickly. Up until 1994 when sledge dogs were removed from Antarctica, a second board was used to record the location of the dogs outside on the spans or in cages close the station. Every day the tag belonging to station's chef was mysteriously moved from his allocated bunk and hung on the hooks denoting the dog spans. As the chef was somewhat hirsute, the perpetrator of this daily niggle found this more than a little amusing. Once could have been forgiven, but the daily switch continued long after it could be considered a joke, and it caused the victim more than a little frustration and embarrassment. The perpetrator was unpleasant, occasionally dangerous, and widely disliked. Unsurprisingly, he never returned to Antarctica while his victim came back many times. His extraordinary service was eventually recognised with the Polar Medal and an important individual prize.

In 1989, I was with one of my greatest BAS friends drinking a wildly expensive gin and tonic in the Taj Lake Palace Hotel bar on the lake in Udaipur, India. My friend and I worked in different departments but found each other through a mutual love of sailing. For a couple of years, we shared a house as well as much of our free time. On this occasion, we were travelling through India on our way up to Pahalgam in the foothills of the Himalayas where we would spend a few days walking and taking photographs.

I remember the lights of the city reflected in the lake, and the ease of the old-world opulence around us. The hotel had been a key location in the 1983 James Bond film 'Octopussy'. Despite it being among the worst films in the Bond franchise, and the fact that drinks were costing more than our accommodation, we were keen to visit.

My friend decided this was as good a time as any to take me into his confidence. Before ordering a second round he announced he was gay. I was not shocked by the fact itself, but I had no inclination that it was something he was ever likely to say. On reflection, he had never feigned any overtly heterosexual behaviour, but to me he was as far from any gay stereotype as I could imagine. He was a great adventurer and had already led several ambitious traverses for BAS. We had climbed together and competed in several long-distance races on a yacht owned by his brother-in-law. We were brothers in arms, and I had no inclination of his secret.

His second revelation was just as surprising. Although he seemed to have a promising science career at BAS, he told me that he had decided to leave and take a job in the city. He simply could not see a future for himself in BAS. On reflection I think he made the right decision. It was a huge sadness to me, but I suspect he felt that BAS was not ready for anyone to be openly gay. We kept in contact for years. He lived in London and then spent a peripatetic decade in the far east. Eventually, we lost contact, but I know that he settled down and had a family.

For women who wanted to pursue a career in Antarctic science it was a frustrating time. Angry at BAS's continued refusal to allow women in Antarctica some ignored the stated requirements and applied anyway. None of them made it beyond the initial sift. I remain friends with one of those women. She was studying for a PhD in Cambridge and was certainly more academically qualified than me. With her PhD completed, she left the UK for New Zealand where no such gender restrictions persisted. There, she thrived and had a stellar career in research, university administration and as an expert advisor to the New Zealand government. Refusing to employ her was a loss to BAS and the UK, and I have always carried a little guilt that I won my place through demographics rather than through merit and although the resistance to employing women to go south with BAS was a hangover from an earlier age, it surprises me that it persisted so long.

Although BAS still has a long way to go, in terms of ethnic diversity, I am proud that it is no longer a place exclusively for people like me. Today BAS celebrates diversity. It works hard to promote and enhance Antarctic science opportunities to under-represented groups, including women, minoritised ethnic groups in the UK, LGBTQ+ and disabled people. It brings together extraordinary, talented and unique people regardless of gender, sexuality or background.

You can find out more about Antarctica and BAS from the website www.bas.ac.uk.

Tipping points and state changes

There are many long-standing questions in science. Perhaps, the most famous is that of the composition of dark matter, and its sister dark energy, which together may account for up to 95% of the mass of the universe. Despite the existence of dark matter being postulated in the 1930s as an explanation for the apparently anomalous patterns of rotation seen within far-off galaxies, there is still no consensus as to what dark matter is made of or how it came into existence.

In glaciology we question how the West Antarctic Ice Sheet (WAIS) - a complex part of the Earth System will behave in a fundamentally different state from the one we observe today or the one we can see existed in the relatively recent past.

If WAIS begins to collapse, it will be because processes, that are barely significant today, have become dominant and are driving ice-sheet retreat. At the same time, it is possible that other new processes will be significant in opposing ice-sheet retreat.

In our case, the retreat of the ice sheet will most likely be driven by the Marine Ice-Cliff Instability (MICI), the enhanced rate of iceberg calving that occurs where ice cliffs higher than around 100 m enter the sea. Some early modelling studies[45] envisaged this process of exporting ice from the ice sheet ice into the ocean as one that could grow without limit. However, it is inevitable that at some stage, a new process will become significant. Perhaps the coastal seas themselves will become clogged with ice and unable to move newly formed icebergs into open water, or perhaps there are factors in the fracture of the ice that will eventually limit calving[46].

If neither of these processes is understood in any detail, and each is not sufficiently common at present to be studied, how will we determine the likely, or even the maximum, rate of ice-sheet retreat? We might be able to formulate a mathematical description of the processes, and even incorporate them into a numerical model, but where are we to find the observations that would allow us to test those models – nowhere that I can see!

---

45 - DeConto, R., Pollard, D. (2016). Nature 531, 591–597.

46 - Basis has already alluded to one possible mechanism. Basis, J. et al. (2021). Science 372, 1342-1344.

I do not have an answer to this conundrum, but it is evident that during a collapse of WAIS there will be a fundamentally different state to the one we observe today. It will be a state that cannot be treated simply as a minor 'perturbation'. The models we currently have, verified using contemporary observations, will not be appropriate. To avoid the need for parameterisations, we might build models based on lower-level physics, but nonetheless, without some verification it will be hard to reassure ourselves that these are working properly.

To my assessment, this reasoning applies not just to future WAIS collapse, but also to a range of so-called 'tipping points' in the Earth System[47]. Such tipping points have previously been characterised in terms of changes that once exceed cannot be reversed. I prefer to think of them in terms of state changes, in which the prevailing condition alters so significantly, that the system enters a very different and consequently unpredictable dynamic.

For example, the release of methane from melting Arctic permafrost, would constitute a fundamental state-change under which we simply do not have the insight to predict which processes will become dominant. The balance would eventually lie between positive and negative feedback.

In a way, this line of reasoning is equivalent in outcome, if not intent, to saying that the Earth System while it might be complex today, has the potential to grow vastly more complex in coming decades and centuries. There is the potential to change way beyond the predictive capacity of science today by entering any one of a multitude of states currently beyond even conjecture.

In short, Gaia may turn out to be fickle mistress. Once irked, she may change dramatically in her mood and her hospitality. In her grand scheme, she will care nought whether our irritating species survives; our Earth mother may simply utter a cosmic 'good riddance' and change beyond our recognition or ability to cope.

I reach not for the language of apocalypse, my state changes are subtle and might pass unnoticed when they occur, especially in far corners of the planet such as Antarctica. Or, in processes of little interest to many except us devotees. But their impact may be sufficient to undermine quite completely the conditions on which our weak and habituated species has grown to depend. Afterall, in terms of the last 20 thousand years a global sea-level rise of a few metres per century is entirely trivial, worse has happened and persisted

[47] - Several global tipping points were identified and explored by Tim Lenton and colleagues, each has the potential to change our planet in profound ways. They include: the release of methane from Arctic permafrost, the ecological collapse of the Amazonian rainforest, the large-scale reorganisation of ocean currents.

for millennia, and it is only human society that would see it in terms of calamity. In this, as with so many others, we have burned our resilience for short-term profit. We have none to blame but ourselves that a bit of sea-level rise in the next couple of hundred years will upset our precious coastal planning or drown a long list of coastal megacities, and with it royally bugger several tens of millions of lives.

Are the state-changes I describe here simply 'tipping points' repackaged? I think not! Tipping points as usually discussed are single, or at least simple thresholds, beyond which things change and from which return is unlikely. However, it seems to me that there is an implication that the future beyond a tipping point is to some degree predictable and stable, and all that is required of science is to predict the threshold at which they will occur and tell society how much longer we can continue without risk.

For me, what is key is the uncertainty that state-changes imply. A state-change may be reached by exceedance of some simple threshold, but the outcome of a state-change is, with our current skill, inherently unpredictable. All we can truly say is that once things have changed, the future will be unpredictably different.

Scientists might be tasked with defining and seeking to observe early indicators that a state-change has begun, or better still, is imminent, but beyond that it is unlikely that we will do a good job in quantification of the outcome. In West Antarctica, the acceleration of ice-loss, and the growth of high ice cliffs may be a sign that collapse has begun. Indeed, both changes have been observed, but how much and how fast the collapse might occur, is beyond us. In honesty this is likely to remain beyond us in a meaningful sense for the foreseeable future of science.

These are problems, at which we could throw any amount of computing resources, the problem is not one of bandwidth, rather it is a lack of any reliable way to know when we have the correct answer. There is simply no natural analogue, no observable data we can use to test our models and without that, how will we ever reach a meaningful consensus from which we can move forward?

On Geoengineering

"We need our reason to teach us today that we are not, that we must not try to be, the lords of all we survey. We are not the lords, we are the Lord's creatures, the trustees of this planet, charged today with preserving life itself—preserving life with all its mystery and all its wonder. May we all be equal to that task."

Margaret Thatcher (November 8th, 1989) - Speech to United Nations General Assembly

Even for UK, let alone the rest of the world, there is a strong likelihood that the trajectory of carbon reductions will fall short of that required to reach net-zero by 2050. Reaching this target will be costly, and ultimately painful. It is inevitable that folk will become more resistant to the tax and legal mechanisms put in place to nudge individuals and commerce towards the expensive and annoying measures that will be required to address the most deeply embedded carbon emissions. Voices will be raised and sound more and more persuasive that exploring and even taking a chance on any 'solution', might relieve the pressure buying us valuable decades.

Some of those voices will favour technological fixes and so-called 'geoengineering solutions'. These are a class of engineering 'techno-fixes' through which we seek to modify the planet, with sufficient strength that the impact of climate change is reduced either locally or globally.

'Geoengineering' is a broad church of techno-fixes that share a common approach of engineering the environment through active intervention on a globally significant scale. A few of the softer geoengineering approaches make a great deal of sense, for example planting trees in industrially deforested areas (e.g., the Amazon rainforest) and perhaps planting new forests in naturally treeless zones. However, many of the schemes proposed are barking mad; for example, deploying vast mirrors into orbits around the Earth that keep them fixed between us and the Sun and keep the planet in perpetual partial shade. But there are geoengineering options that sit in the middle, not obviously mad, but certainly complex, costly and with the potential to cause unforeseen consequences.

Take, for example, the several schemes that have been proposed to 'whiten' the Arctic, counteracting the ongoing effects of reduced sea ice and snow cover, and by increasing the reflection of sunlight back to space, reducing high latitude warming. I know of no credible plan as to how this might be achieved, but it has been suggested it might include floating granules or foam distributed over the Arctic Ocean. However, supposing we accept the crippling cost and direct environmental impact of such a scheme, and the likelihood that it would be successful[48], there are likely serious side-effects of any attempt to whiten the Arctic. For example, the is a potential that by increasing the temperature difference between the Arctic and the Tropics will simply add energy to weather systems and deliver mid-latitudes more and stronger extreme weather events, with the consequence that far-distant weather patterns could be disrupted.

From our current perspective in the early-2020s, it might seem that geoengineering approaches will always be too costly and too uncertain to warrant serious consideration, but my concern is that perspectives will change.

Imagine discussions at COP40, due in a mere fourteen years' time. Delegates from every nation are gathered to discuss the depressing global progress towards net-zero. We are already half-way to 2050, and carbon emissions are tracking well above what is required to reach our target. There is rising clammer of voices of techno-saviours, entrepreneurs and big businesses, who argue that geoengineering fixes can give us an extra couple of decades before we need to make the most painful cuts. Furthermore, the same entrepreneurs and businesses are offering vast research budgets to develop proof-of-concept projects. Given the urgency to find a solution, and the sheer strength of business muscle that could be brought to bear, even our heroes, the voices of reason might struggle to be heard. Our governments seeking to make progress during their brief tenures, whilst at the same time trying to avoid upsetting the electorate might well see investment in geoengineering as an effective way to play down the clock. It might take only one leader – one with Messianic tendencies and an eye on being saviour of the world – to champion the cause, and the stage could be set for a global experiment with appalling and ruinous consequences. This much seems inevitable, but also unstoppable, my concern is how that debate plays out, and to ensure that the science community's voice is ready to be heard.

It is hard to evidence concretely, but I have seen for many years an ingrained antipathy on the part of the UK's research funders to address anything so shabby as geoengineering, and a similar reluctance in the scientific community to get involved.

---

[48] - Several studies suggest that whitening of the Arctic would not have the desired impact on climate beyond the Arctic. See, for example: Ivana Cvijanovic et al., Impacts of ocean albedo alteration on Arctic-sea ice restoration and Northern Hemisphere climate. Environmental Research Letters, 2015; 10 (4)

My own experience in this area is still raw. A few years back, one of the best young researchers I have known pitched her idea for a fellowship proposal to me. She was interested in the co-impacts of a particular geoengineering solution whose aim was to increase cloud-cover over the Southern Ocean. Her motivation was curiosity, but she clearly came from an open-minded but sceptical viewpoint. I judged the project to be one of the strongest I had seen, and with her stellar CV, it made a very strong case for a fellowship proposal. Several of us worked hard to refine and polish her proposal, but it was rejected twice and the researcher in question has since left science, at least for the present. I am sorry, if I gave her bad advice, and wish now I had told her to stick to something less provocative.

For the academics themselves there is a question of motivation. There are almost no role-models of environmental scientists who have engaged in evaluating geoengineering and the few there are, happily present themselves as anti-establishment figures. Few if any scientific prizes will be won by scientists engaged in research on geoengineering schemes. And finally, as I have said, many directors of science believe that any research that might, even inadvertently, give credibility to geoengineering research should be strangled at the proposal stage.

If we are as a nation and a rational society going to continue to lead the world towards net-zero, we must use our fear and emotion to urge us forward with energy and clarity of purpose, but never allow that emotion to cloud our judgement, or blunt the keenness of our critical analysis of cost, benefit, and impact. In future, we will need independent research about geoengineering, its benefits, and more crucially its pitfalls, to inform public debate. Researchers themselves must engage, and communicators strive for logic and not emotion. The clock is running down, and rather soon we will need to make some decisions.

During his 36-year career at British Antarctic Survey (BAS) Professor David Glyn Vaughan, OBE became recognised nationally and internationally as a leading expert on understanding the response of polar ice sheets to climate change and the implications for society.

Born in Cyprus while his father was working for the Met Office in Libya, David went to school in England where he shone at maths, science and exams. He went on to read Natural Sciences at Churchill College, Cambridge, followed by a Masters in geophysics at Durham University.

Over a period of 15 years, he served as co-ordinating lead author in two rounds of the Intergovernmental Panel on Climate Change (IPCC) assessment reports. He was responsible for identifying the policy-relevant issues and negotiating the acceptance of key findings by high-level policymakers. His enthusiasm for science communication enabled wider publics to understand why the Polar Regions are crucial for planet Earth.

As the former Director of Science at BAS, David was the UK lead for the International Thwaites Glacier Programme. He retired from BAS in July 2021.

Bury me with my glasses. If I should find myself at the gates of heaven, I will need them to read the small print.

Printed in the United States
by Baker & Taylor Publisher Services